Willmore Energy and Willmore Conjecture

MONOGRAPHS AND RESEARCH NOTES IN MATHEMATICS

Published Titles

Actions and Invariants of Algebraic Groups, Second Edition, Walter Ferrer Santos and Alvaro Rittatore

Analytical Methods for Kolmogorov Equations, Second Edition, Luca Lorenzi

Application of Fuzzy Logic to Social Choice Theory, John N. Mordeson, Davender S. Malik and Terry D. Clark

Blow-up Patterns for Higher-Order: Nonlinear Parabolic, Hyperbolic Dispersion and Schrödinger Equations, Victor A. Galaktionov, Enzo L. Mitidieri, and Stanislav Pohozaev

Bounds for Determinants of Linear Operators and Their Applications, Michael Gil′

Complex Analysis: Conformal Inequalities and the Bieberbach Conjecture, Prem K. Kythe

Computation with Linear Algebraic Groups, Willem Adriaan de Graaf

Computational Aspects of Polynomial Identities: Volume l, Kemer's Theorems, 2nd Edition Alexei Kanel-Belov, Yakov Karasik, and Louis Halle Rowen

A Concise Introduction to Geometric Numerical Integration, Fernando Casas and Sergio Blanes

Cremona Groups and Icosahedron, Ivan Cheltsov and Constantin Shramov

Delay Differential Evolutions Subjected to Nonlocal Initial Conditions Monica-Dana Burlică, Mihai Necula, Daniela Roşu, and Ioan I. Vrabie

Diagram Genus, Generators, and Applications, Alexander Stoimenow

Difference Equations: Theory, Applications and Advanced Topics, Third Edition Ronald E. Mickens

Dictionary of Inequalities, Second Edition, Peter Bullen

Elements of Quasigroup Theory and Applications, Victor Shcherbacov

Finite Element Methods for Eigenvalue Problems, Jiguang Sun and Aihui Zhou

Introduction to Abelian Model Structures and Gorenstein Homological Dimensions Marco A. Pérez

Iterative Methods without Inversion, Anatoly Galperin

Iterative Optimization in Inverse Problems, Charles L. Byrne

Line Integral Methods for Conservative Problems, Luigi Brugnano and Felice Iavernaro

Lineability: The Search for Linearity in Mathematics, Richard M. Aron, Luis Bernal González, Daniel M. Pellegrino, and Juan B. Seoane Sepúlveda

Modeling and Inverse Problems in the Presence of Uncertainty, H. T. Banks, Shuhua Hu, and W. Clayton Thompson

Published Titles Continued

Monomial Algebras, Second Edition, Rafael H. Villarreal

Noncommutative Deformation Theory, Eivind Eriksen, Olav Arnfinn Laudal, and Arvid Siqveland

Nonlinear Functional Analysis in Banach Spaces and Banach Algebras: Fixed Point Theory Under Weak Topology for Nonlinear Operators and Block Operator Matrices with Applications, Aref Jeribi and Bilel Krichen

Optimization and Differentiation, Simon Serovajsky

Partial Differential Equations with Variable Exponents: Variational Methods and Qualitative Analysis, Vicenţiu D. Rădulescu and Dušan D. Repovš

A Practical Guide to Geometric Regulation for Distributed Parameter Systems Eugenio Aulisa and David Gilliam

Reconstruction from Integral Data, Victor Palamodov

Signal Processing: A Mathematical Approach, *Second Edition*, Charles L. Byrne

Sinusoids: Theory and Technological Applications, Prem K. Kythe

Special Integrals of Gradshteyn and Ryzhik: the Proofs – Volume I, Victor H. Moll

Special Integrals of Gradshteyn and Ryzhik: the Proofs – Volume II, Victor H. Moll

Spectral and Scattering Theory for Second-Order Partial Differential Operators, Kiyoshi Mochizuki

Stochastic Cauchy Problems in Infinite Dimensions: Generalized and Regularized Solutions, Irina V. Melnikova

Submanifolds and Holonomy, Second Edition, Jürgen Berndt, Sergio Console, and Carlos Enrique Olmos

Symmetry and Quantum Mechanics, Scott Corry

The Truth Value Algebra of Type-2 Fuzzy Sets: Order Convolutions of Functions on the Unit Interval, John Harding, Carol Walker, and Elbert Walker

Variational-Hemivariational Inequalities with Applications, Mircea Sofonea and Stanislaw Migórski

Willmore Energy and Willmore Conjecture, Magdalena D. Toda

Forthcoming Titles

Groups, Designs, and Linear Algebra, Donald L. Kreher

Handbook of the Tutte Polynomial, Joanna Anthony Ellis-Monaghan and Iain Moffat

Microlocal Analysis on R^n and on NonCompact Manifolds, Sandro Coriasco

Practical Guide to Geometric Regulation for Distributed Parameter Systems, Eugenio Aulisa and David S. Gilliam

MONOGRAPHS AND RESEARCH NOTES IN MATHEMATICS

Willmore Energy and Willmore Conjecture

Edited by
Magdalena D. Toda

CRC Press is an imprint of the
Taylor & Francis Group, an **informa** business
A CHAPMAN & HALL BOOK

CRC Press
Taylor & Francis Group
6000 Broken Sound Parkway NW, Suite 300
Boca Raton, FL 33487-2742

Printed on acid-free paper
Version Date: 20170831

International Standard Book Number-13: 978-1-4987-4463-8 (Hardback)

Visit the Taylor & Francis Web site at
http://www.taylorandfrancis.com

and the CRC Press Web site at
http://www.crcpress.com

Dedication

To my son and my husband

Contents

Preface

Mathematics is a living tree. Its branches grow and ramify continuously. The study of Willmore energy had originally emerged from differential geometry and solid mechanics. Its neighboring branches include mathematical physics, calculus of variations, algebraic geometry, partial differential equations and measure theory.

This monograph is primarily dedicated to the memory of Thomas James Willmore (1919 – 2005), the father of Willmore surfaces, whose academic presence is still alive at the University of Durham, where he spent a lengthy period of his career. It is a modest tribute, but I would hope it will become as inspiring as the sculpture by Peter Sales, representing a 4-lobed Willmore torus, which was unveiled at the University of Durham, in March 2012.

Secondly, the monograph is a tribute of gratitude to Fernando Codá Marques and André Arroja Neves, the mathematicians who proved Willmore's famous conjecture stated in 1965. Their proof, announced in 2012, was published in 2014 in Annals of Mathematics (Chapter 1, [10]).

The long-term goal of this monograph is not that of reproducing famous results that have already been published and widely disseminated, but that of presenting some new directions, developments and open problems in the field of contemporary Willmore energy and Willmore surfaces. It is meant to answer questions like:

- What is new in Willmore theory?
- Are there any new Willmore conjectures and open problems?

Mentioning all the people who contributed to the field of Willmore energy and Willmore surfaces would be a foolish task, one that is impossible to achieve. I am therefore apologizing in advance to everyone who was not mentioned in this work, in spite of significantly contributing the field.

The authors of this monograph, just like all who worked in the field of Willmore energy, come from many different branches of mathematics.

Although S.D. Poisson and M.S. Germain are those who first introduced the notion of Willmore (bending) energy, the righteous founders of the corresponding surface theory are T. Willmore (Chapter 1, [17]) and W. Blaschke (Chapter 1, [4]) - whose work was performed independently. As early as 1920s, Blasche had extensively studied *conformal minimal surfaces* - which represent Willmore surfaces.

Some remarkable contributions to the geometric theory of Willmore energy and Willmore surfaces field were brought by R. Bryant and P. Griffiths.

Many other significant contributions to the theoretical and computational studies of Willmore surfaces, constrained Willmore surfaces, and associated flows, were brought by: B. Ammann, D. Berdinsky, M. Bergner, Y. Bernard, A.I. Bobenko, C. Bohle, D. Brander, F. Burstall, J. Chen, A. Dall'Acqua, K. Deckelnick, P. Djondjorov, P. Dondl, J. Dorfmeister, M. Droske, G. Dziuk, N. Ejiri, C.M. Elliott, S. Froelich, C. Graham, H. Grunau, M. Gürses, M. Hadzhilazova, F. Hélein, L. Heller, U. Hertrich-Jeromin, J. Hirsch, P. Hornung, Y. Hu, J. Inoguchi, R. Jacob, L. Ji, J. Katz, A.I. Kholodenko, B. G. Konopelchenko, M. Kilian, E. Kuwert, T. Lamm, P. Laurain, K. Leschke, H. Li, J. Li, Y. Li, F. Link, Y. Luo, X. Ma, A. Malchiodi, S. Masnou, R. Mazzeo, J. Metzger, I. Mladenov, A. Mondino, G. Nardi, C. Ndiaye, V.V. Nesterenko, F. Pedit, R. Perl, G. Peters, U. Pinkall, P. Pozzi, A. Quintino, T. Rivière, M. Röger, W. Rossman, M. Rumpf, S. Santos, R. Schätzle, F. Schieweck, M.U. Schmidt, N. Schmitt, P. Schröder, H. Schumacher, Y. Shao, L. Simon, G. Simonett, I. Sterling, J.M. Sullivan, J. Sun, I.A. Taimanov, V. Vassilev, C. Wang, P. Wang, G. Wheeler, M. Xu, Y. Xu, W. Yan, to name just a few.

Software Notes

In this work we use the software MATLAB®

```
The MathWorks, Inc.
3 Apple Hill Drive
Natick, MA 01760-2098 USA
Tel: +1-508-647-7000,  +1-Fax: 508-647-7001
E-mail: info@mathworks.com Web: www.mathworks.com
```

We also use the software Mathematica®

```
WOLFRAM
100 Trade Center Drive
Champaign, IL 61820-7237 USA
Tel: +1-217-398-0700,  Fax: +1-217-398-0747
Web: www.wolfram.com/company/contact/?source=nav
```

Acknowledgments

I would like to express my thanks to all the contributing authors of this monograph. It has been an honor and a pleasure corresponding with them and learning from their expertise.

No work is perfect: I apologize in advance for any types of errors and/or inconsistencies that may exist in this monograph, and I am truly grateful to all the contributors and readers for their suggestions and criticism.

Last but not least, I would like to thank all those faculty members at Texas Tech University who supported me in the process of working on the monograph. This work has been edited while I have been a department chair, besides being actively involved in teaching, research and service. I appreciate everyone's support, patience and encouragement.

A very special kind of gratitude goes to my husband and colleague, Eugenio Aulisa, for his editorial help, and to my 12 year old son, Lorenzo Aulisa, for his patience with a very busy mother.

Chapter 1

Willmore Energy: Brief Introduction and Survey

M. D. Toda

CONTENTS

The Willmore energy is a type of energy that is studied in differential geometry, physics and mechanics. It was first introduced by Siméon-Denis Poisson and Marie-Sophie Germain, independently, at the beginning of the nineteenth century, but its complete formalism was due to Thomas Willmore. In essence, this type of energy quantitatively measures the deviation of a surface from local sphericity. It is important to note that there is some ambiguity in the literature between the terms *bending energy* and *Willmore energy*, and different sources provide different definitions.

Due to historic and pragmatic reasons alike, we use the following terminology:

Definition 1.1. *Let M be a smooth, orientable surface immersed in $\mathbb{R}^3$. We define the Willmore energy functional as*

$$W(M) = \int_M H^2\, dS, \tag{1.1}$$

where the term dS is the area element with respect to the induced metric, and H is the mean curvature.

Throughout the mathematical literature, many authors (e.g., [5]) define the Willmore energy functional as

$$\tilde{W}(M) = \frac{1}{4}\int_M (k_1 - k_2)^2\, dS = \int_M (H^2 - K)\, dS \tag{1.2}$$

which is what we usually call *bending energy*, where H, K and $k_i, i = 1, 2$

respectively represent the mean curvature, Gaussian curvature and principal curvatures of the surface M. Marie-Sophie Germain had actually proposed, as *the bending energy of a thin plate*, the integral with respect to the surface area of a symmetric function of the principal curvatures (see [8]). Note that the Willmore energy is just one instance of a multitude of bending energies. In this sense, Sophie Germain's definition of bending energy is a very natural form of *generalized Willmore energy.*

$\tilde{W}$ presents the advantage of vanishing exactly at the umbilic points of the immersion.

Clearly, for a closed surface, by the Gauss-Bonnet theorem, the integral of the Gaussian curvature represents the quantity $2\pi\chi(M)$, where $\chi(M)$ is the Euler characteristic of the surface. Therefore, $\tilde{W}$ has the same critical points as the original functional (1.1). Therefore, the study of (1.1) is equivalent to the study of (1.2).

Theorem 1.2. *Let M be a closed, orientable surface immersed into the Euclidean 3-space via $f : M \to \mathbb{R}^3$. Then*

$$W(M) = \int_M H^2 \, dS \geq 4\pi. \tag{1.3}$$

Moreover, $W(M) = 4\pi$ if and only if M is the round sphere embedded in $\mathbb{R}^3$.

This theorem (published in [18]) was easily proved by Willmore and can be left an an exercise to the talented graduate student. Furthermore, the equality holds if and only if $k_1 = k_2$ at every point. Further calculations on a few examples of closed surfaces had suggested that there could exist a better bound than (1.3) if the surface has genus $g(M) > 0$.

Further, Willmore proved the following theorem (see [18]).

Theorem 1.3. *Let M be a torus embedded in $\mathbb{R}^3$ as a tube of constant circular cross-section. Then*

$$\int_M H^2 \, dS \geq 2\pi^2, \tag{1.4}$$

the equality holding if and only if the generating curve is a circle and the ratio of the radii is $1/\sqrt{2}$.

In 1965, Willmore proposed to extend the previous result to any torus by formulating his famous conjecture:

"For every smooth torus M that is immersed in $\mathbb{R}^3$, $W(M) \geq 2\pi^2$".

This was a challenge for mathematicians for over 45 years. In 2012, the conjecture was proved by Fernando Codá Marques and André Arroja Neves, by using the famous Almgren-Pitts min-max theory.

The reader is strongly encouraged to read the theorem and its beautiful proof in [11]:

Theorem 1.4. *Every embedded compact surface M in $\mathbb{R}^3$ with positive genus satisfies $W(M) \geq 2\pi^2$. Up to rigid motions, the equality holds only for stereographic projections of the Clifford torus.*

It is important to remark that many outstanding mathematicians have used this type of technique for different proofs, including but not limited to: M. L. Gromov, R. Schoen, S.-T. Yau, F. C. Marques, A. A. Neves, I. Agol, in addition to its founders (F. Almgren and J. Pitts).

Let us go five decades back in time. Thomas Willmore himself had showed that the energy $W(M)$ of the immersion M is invariant under conformal transformations in Euclidean space. He proved the following theorem using the fact that any conformal transformation of $\mathbb{R}^3$ can be decomposed into a product of similarity transformations and inversions (see again [18]):

Theorem 1.5. *Let $f : M \to \mathbb{R}^3$ be a smooth surface immersion of a compact orientable surface into $\mathbb{R}^3$. Let*

$$W(M) = \int_M H^2 \, dS.$$

Then, $W(M)$ is invariant under conformal transformations of $\mathbb{R}^3$ [18].

This result was already known to Blaschke [4].

1.1 The Euler-Lagrange equation of the Willmore functional

The history of this Euler-Lagrange equation corresponding to the Willmore energy functional is highly unusual. In 1923, it appeared in the Ph.D. dissertation of Gerhard Thomsen ([15]), who was a student of Wilhelm Blaschke. Independently, Konrad Voss also obtained this equation in the 1950s, but he never published his work. On the other hand, its complete formalism was due to Willmore.

Willmore himself proved this Euler-Lagrange equation using the standard techniques of calculus of variations. He considered the normal variation of the immersion along with Gauss-Weingarten equations and Green's second identity on closed surfaces [18].

The following definition of Willmore immersion can be found in [14]:

Definition 1.6. *Let $\mathbf{s} : M \to \mathbb{R}^n$ be a smooth surface immersion such that $W(\mathbf{s}) < \infty$. The map $\mathbf{s}$ is a critical point for W if*

$$\forall \mathbf{w} \in C^\infty(M, \mathbb{R}^n), \qquad \frac{d}{dt} W(\mathbf{s} + t\mathbf{w}) \Big|_{t=0} = 0. \tag{1.5}$$

Such an immersion is called a ***Willmore*** *surface.*

Theorem 1.7. *A smooth immersion* $\mathbf{r} : M \to \mathbb{R}^3$ *is Willmore if and only if it solves the equation*

$$\Delta_g H + 2H(H^2 - K) = 0, \tag{1.6}$$

where Δ_g *is the Laplace Beltrami operator which depends on the natural metric.*

At this moment, the study of Willmore energy minimizers is under continuous and effervescent development.

As we all learned between 2012 and 2014, the global minimizer of the Willmore energy in the class of tori is the Clifford torus (see [11]).

Recently, it was proved in [12] that the Clifford torus minimizes the Willmore energy in an open neighbourhood of its conformal class, in arbitrary codimension, where the neighbourhood may depend on the codimension.

In 2014, M. Kilian, M.U. Schmidt and N. Schmitt published the proof of a remarkable new result, namely that: "amongst the equivariant constant mean curvature tori in the 3-sphere, the Clifford torus is the only local minimum of the Willmore energy. All other equivariant minimal tori in the 3-sphere are local maxima of the Willmore energy" (see [10]).

There is a significant amount of research devoted to generalized Willmore energies, whose applications appear in physics and biophysics.

1.2 Extensions and generalizations of Willmore energy

Definition 1.8. *The terminology used in this section closely follows that from T. Paragoda's Ph.D. dissertation ([13], 2016).*

The standard generalized Willmore energy functional associated to an elastic surface (membrane) M *immersed in* $\mathbb{R}^3$ *is given by*

$$W(M) = \int_M (cH^2 + \mu)\, dS, \tag{1.7}$$

where $c = 2k_c$ *is the double of the usual bending rigidity* k_c *and* μ *is the surface tension. The term* dS *is the area element with respect to the induced metric.*

The corresponding Euler-Lagrange equation of (1.7) is given by

$$\Delta_g H + 2H(H^2 - K - \epsilon) = 0, \tag{1.8}$$

where $\epsilon = \mu/c$.

Note that this represents a particular case of energies of the type

$$\int_M (\alpha H^2 + \beta K + \gamma) dS$$

(see, for example, [2]). The functional (1.7) has particular relevance for membranes with bending rigidity and surface tension, where other rigidities can be neglected [2]. The generalizations of the Willmore energy include the famous *Helfrich energy for lipid bilayers*, whose study was much further developed by Helfrich's collaborators, physicists Z. C. Tu and Z. C. Ou-Yang, through several works, of which we would only like to mention [16] and [17].

It is important to remark that other scientists, in various circumstances, have used the term *generalized Willmore energy* for a more comprehensive type of energy.

One such example is provided by *free energy of an open lipid bilayer with boundary*

$$\oint \varepsilon(H, K, t)dS,$$

where ε is a smooth function of a specific real parameter t, as well as H and K (the mean curvature and Gaussian curvature of an elastic surface, respectively). For specifics, please consult [16] and [17]. Remarkably, such a model can be applied to a large variety of elastic surfaces. The Euler-Lagrange equations that correspond to the free energy can be expressed as follows [17]:

$$\int (\partial\varepsilon/\partial t)dS = 0 \tag{1.9}$$

$$(\nabla^2/2 + 2H^2 - K)\partial\varepsilon/\partial H + (\nabla \cdot \tilde{\nabla} + 2KH)\partial\varepsilon/\partial K - 2H\varepsilon = 0 \tag{1.10}$$

The definition of the operator $\tilde{\nabla}$ can be found in the appendix of [16].

Interesting developments occur from studying a generalized Willmore flow, which is the L^2-gradient flow corresponding to a generalized Willmore energy.

The Willmore flow has had an important role in digital geometry processing, geometry modeling and physical simulation. In the literature, Droske and Rumpf derived a level set formulation for Willmore flow and used level set equations [7].

In terms of discrete surfaces, a most remarkable contribution is [5], where A.I. Bobenko and P. Schröder studied the discrete Willmore energy and its flow, and derived the relevant gradient expressions including a linearization (approximation of the Hessian), as being required for nonlinear numerical solvers. In 2008, [3] presented a parametric approximation of Willmore flow and related geometric evolution equations. They provided various numerical simulations for energies appearing in the modeling of biological cell membranes. Some authors [6, 9] studied the error estimates for the Willmore flow of graphs along with numerical simulations.

As for the *generalized Willmore flows*, there are no significant studies we are aware of, except those from [1] and [13].

In [1] and [13], a novel numerical scheme for solving a *generalized Willmore flow equation* has been presented. The team formulated the geometric evolution equations as coupled systems of nonlinear PDEs where the unknowns are the profile and the weighted mean curvature. To solve these coupled systems, the authors made use of Automatic Differentiation (AD) techniques to compute the Jacobian in the Newton linearization of the finite element weak formulation. They have tested the accuracy of the algorithm by providing nontrivial steady-state numerical solutions of the generalized Willmore flow equation. The work can be extended in several directions. The numerical scheme and the implementation thereby presented can be applied to time-dependent problems and to conformal immersions in $\mathbb{R}^3$. These studies are expected to bring new interesting results and future high-speed computational applications of generalized Willmore surfaces and flows.

Bibliography

[1] Athukoralage B., Aulisa E., Bornia G., Paragoda T., and Toda M. *New advances in the study of Generalized Willmore surfaces and flow.* Seventeenth International Conference on Geometry, Integrabiity and Quantization, pages 1–11, 2015.

[2] Athukoralage B. and Toda M. *Geometric models for secondary structures in proteins.* AIP Proceedings - American Institute of Physics, 1558(883), 2013.

[3] Barrett J. W., Garcke H., and Nurberg R. *Parametric approximation of Willmore flow and related geometric evolution equations.* SIAM J. Sci. Comput., 31(1):225–253, 2008.

[4] Blaschke W. *Einfuhrung in die Differentialgeometrie.* Berlin, Springer, 1950.

[5] Bobenko A. I. and Schröder P. *Discrete Willmore flow.* Eurographics Symposium on Geometry Processing, pages 101–110, 2005.

[6] Deckelnick K., Katz J., and Schieweck F. *A C^1-finite element method for the Willmore flow of two dimensional graphs.* Mathematics of computation, 84(296):2617–2643, 2015.

[7] Droske M. and Rumpf M. *A level set formulation for Willmore flow.* Interfaces free boundaries, pages 361–378, 2004.

[8] Germain S. *Recherches sur la théorie des surfaces élastiques*, V. Courcier, Paris (1821).

[9] Ji L. and Xu Y. *Optimal error estimates of the local discontinuous Galerkin method for Willmore flow of graphs on cartesian meshes.* International Journal of Numerical Analysis and Modeling, 8(2):252–283, 2011.

[10] Kilian M., Schmidt M.U., and Schmitt N. *On stability of equivariant minimal tori in the 3-sphere.* Geometry and Physics, 85:171–176, 2014.

[11] Marques F. and Neves A. *Min-max theory and the Willmore conjecture.* Annals Of Mathematics, 179(2):683–782, 2014.

[12] Ndiaye C.B. and Schätzle R.M. *Explicit conformally constrained Willmore minimizers in arbitrary codimension.* Calc. Var., 51:291–314, 2014.

[13] Paragoda T. *Willmore and generalized Willmore energies in space forms.* TTU Electronic Library of PhD Dissertations, 2016.

[14] Rivière T. *Weak immersions of surfaces with l^2-bounded second fundamental form.* PCMI Graduate Summer School, 2013.

[15] Thomsen G. *Ùber konforme geometrie 1: Grundlagen der konformen flàchentheorie.* Lutcke & Wulff, 1923.

[16] Tu Z.C. and Ou-Yang Z.C. *Geometry theory on the elasticity of biomembranes.* Journal of Physics A: Mathematical and General, 37:11407–11429, 2004.

[17] Tu Z.C. and Ou-Yang Z.C. *Variational problems in elastic theory of biomembranes, smectic-a liquid crystals, and carbon related structures.* Seventh International Conference on Geometry, Inegrability and Quantization, pages 237–248, 2005.

[18] Willmore T. J. *Riemannian Geometry.* Oxford Science Publications. Oxford University Press, 1993.

Chapter 2

Transformations of Generalized Harmonic bundles and Constrained Willmore Surfaces

A. C. Quintino

CONTENTS

Abstract

Willmore surfaces are the extremals of the Willmore functional (possibly under a constraint on the conformal structure). With the characterization of Willmore surfaces by the (possibly *perturbed*) harmonicity of the mean curvature sphere congruence [1, 5, 13, 19], a zero-curvature formulation follows [5]. Deformations on the level of harmonic maps prove to give rise to deformations on the level of surfaces, with the definition of a spectral deformation [5, 8] and of a Bäcklund transformation [9] of Willmore surfaces into new ones, with a Bianchi permutability between the two [9]. This text is dedicated to a self-contained account of the topic, from a conformally-invariant viewpoint, in Darboux's light-cone model of the conformal n-sphere.

2.1 Introduction

Among the classes of Riemannian submanifolds, there is that of *Willmore surfaces*, named after T. Willmore [23] (1965), although the topic was mentioned by Blaschke [1] (1929) and by Thomsen [21] (1923), as a variational problem of optimal geometric realization of a given compact surface in 3-space regarding the minimization of some natural energy.

Early in the nineteenth century, Germain [14, 15] studied elastic surfaces. On her pioneering analysis, she claimed that the elastic force of a thin plate is proportional to its mean curvature, $H = (k_1 + k_2)/2$, for k_1 and k_2 the maximum and minimum curvatures among all intersections of the surface with perpendicular planes, at each point. Since then, the mean curvature remains a key concept in the theory of elasticity.

In modern literature on the elasticity of membranes, a weighted sum of the total mean curvature, the total squared mean curvature and the total Gaussian curvature is considered the elastic energy of a membrane. By neglecting the total mean curvature, by physical considerations, and having in consideration the Gauss-Bonnet Theorem, T. Willmore defined the *Willmore energy* of a compact oriented surface Σ, without boundary, isometrically immersed in $\mathbb{R}^3$ to be

$$\mathcal{W} = \int_\Sigma H^2 dA,$$

averaging the mean curvature square over the surface.

From the perspective of energy extremals, the Willmore functional may be extended to isometric immersions ϕ of compact oriented surfaces Σ in a general Riemannian manifold M of constant sectional curvature by means of half (or any other scale) of the total squared norm of

$$\Pi_0 = \Pi - g_\phi \otimes \mathcal{H},$$

the trace-free part of the second fundamental form Π, for $\mathcal{H} = \frac{1}{2}\mathrm{tr}(\Pi)$, the mean curvature vector, and g_ϕ the metric induced in Σ by ϕ. In fact, given $(X_i)_{i=1,2}$ a local orthonormal frame of $T\Sigma$, the Gauss equation, relating the curvature tensors of Σ and M, establishes, in particular,

$$K - \hat{K} = (\Pi(X_1, X_1), \Pi(X_2, X_2)) - (\Pi(X_1, X_2), \Pi(X_1, X_2)),$$

for K the Gaussian curvature of Σ and $\hat{K}$ the sectional curvature of M, and, therefore,

$$|\Pi_0|^2 = \sum_{i,j} (\Pi_0(X_i, X_j), \Pi_0(X_i, X_j)) = 2(|\mathcal{H}|^2 - K + \hat{K}).$$

Hence, for the particular case of surfaces in $\mathbb{R}^3$, the two functionals share critical points.

Willmore surfaces are the extremals of the Willmore functional. *Constrained Willmore surfaces* appear as the generalization of Willmore surfaces that arises when we consider extremals of the Willmore functional with respect to *infinitesimally conformal variations*, rather than with respect to all variations. The Euler-Lagrange equations include a Lagrange multiplier. Willmore surfaces are the constrained Willmore surfaces admitting the zero multiplier. The zero multiplier is not necessarily the only multiplier for a constrained Willmore surface with no constraint on the conformal structure, though. In fact, the uniqueness of multiplier characterizes [9] non-isothermic constrained Willmore surfaces. Constant mean curvature surfaces in 3-dimensional space-forms are examples of isothermic constrained Willmore surfaces, as proven by J. Richter [18]. A classical result by Thomsen [21] characterizes isothermic Willmore surfaces in 3-space as minimal surfaces in some 3-dimensional space-form.

It is well-known that the Levi-Civita connection is not a conformal invariant. In fact (see, for example, [24, Section 3.12]), under a conformal change $g' = e^{2u}g$ of a metric g on Σ, for some $u \in C^\infty(\Sigma, \mathbb{R})$, the Levi-Civita connections ∇ and ∇' on (Σ, g) and (Σ, g'), respectively, are related by

$$\nabla'_X Y = \nabla_X Y + (Xu)Y + (Yu)X - g(X, Y)(du)^*,$$

for $(du)^*$ the contravariant form of du with respect to g, for all $X, Y \in \Gamma(T\Sigma)$. It follows that, under a conformal change of metric on a Riemannian manifold M, the second fundamental form of an isometric immersion $\phi : \Sigma \to M$ changes according to

$$\Pi'(X, Y) = \Pi(X, Y) - g_\phi(X, Y)\pi_{N_\phi}(\phi^*(du)^*), \tag{2.1}$$

for π_{N_ϕ} the orthogonal projection of the pull-back bundle ϕ^*TM onto the normal bundle $N_\phi = (d\phi(T\Sigma))^\perp$ and $\phi^*(du)^*$ the pull-back by ϕ of $(du)^*$; and, therefore,

$$\mathcal{H}' = e^{-2u\circ\phi}\mathcal{H} - e^{-2u\circ\phi}\pi_{N_\phi}(\phi^*(du)^*),$$

relating the respective mean curvature vectors. Hence, under a conformal change of the metric, the trace-free part of the second fundamental form remains invariant, so that its squared norm and the area element change in an inverse way, leaving the Willmore energy unchanged. In particular, this establishes the class of constrained Willmore surfaces as a conformally-invariant class.

Conformal invariance motivates us to move from Riemannian to conformal geometry. Our study is one of surfaces in n-dimensional space-forms from a conformally-invariant viewpoint. For this, we find a convenient setting in Darboux's light-cone model [12] of the conformal n-sphere, viewing the n-sphere not as the round sphere in the Euclidean space $\mathbb{R}^{n+1}$ but as the celestial sphere in the Lorentzian spacetime $\mathbb{R}^{n+1,1}$.

A manifestly conformally-invariant formulation of the Willmore energy is presented, following the definition presented by Burstall, Ferus, Leschke, Pedit and Pinkall [7], in the quaternionic setting, for the particular case of $n = 4$.

A fundamental construction in conformal geometry of surfaces is the *mean curvature sphere congruence*, the bundle of 2-spheres tangent to the surface and sharing mean curvature vector with it at each point (although the mean curvature vector is not conformally-invariant, under a conformal change of the metric it changes in the same way for the surface and the osculating 2-sphere). From the early twentieth century, with the work of Blaschke [1], the family of mean curvature spheres has been known as the *central sphere congruence*. Nowadays, after Bryant's paper [3], it goes as well by the name *conformal Gauss map*.

A key result by Blaschke [1] ($n = 3$) and, independently, Ejiri [13] and Rigoli [19] (general n) characterizes Willmore surfaces by the harmonicity of the central sphere congruence. The well-developed theory of harmonic maps, and, in particular, the integrable systems approach to these, then applies. The starting point is the fact that, for a map into a Grassmannian, harmonicity amounts to the flatness of a certain family of connections depending on a spectral parameter, according to Uhlenbeck [22]. A zero-curvature characterization of Willmore surfaces follows. This characterization generalizes to constrained Willmore surfaces, as established by Burstall and Calderbank [5].

The zero-curvature representation of the harmonic map equations allows one to deduce two kinds of symmetry: harmonic maps admit a *spectral deformation* [22], by exploiting a scaling freedom in the spectral parameter, and *Bäcklund transformations*, which arise by applying chosen gauge transformations to the family of flat connections, as studied by Terng and Uhlenbeck [20, 22]. Aiming to apply this theory to constrained Willmore surfaces, and in order to address the possibly non-harmonic central sphere congruences of constrained Willmore surfaces, the notion of *perturbed harmonicity* for a map into a Grassmannian is introduced [9]. It applies to the central sphere congruence and it provides a characterization of constrained Willmore surfaces.

A spectral deformation and Bäcklund transformations of perturbed harmonic maps into new ones are defined [9]. Some care is required to see that, when applied to the central sphere congruence of a constrained Willmore surface, each new map still is the central sphere congruence of a surface. Deformations on the level of perturbed harmonic maps prove [9] to give rise to deformations on the level of surfaces, with the definition of a spectral deformation and of Bäcklund transformations of constrained Willmore surfaces into new ones. This spectral deformation of constrained Willmore surfaces coincides, up to reparametrization, with the one presented by Burstall, Pedit and Pinkall [8], in terms of the *Schwarzian derivative* and the *Hopf differential*, later defined by the action of a loop of flat connections, by Burstall and Calderbank [5].

The class of constrained Willmore surfaces is in this way established as a class of surfaces with strong links to the theory of integrable systems, admitting a spectral deformation and a Bäcklund transformation, with a Bianchi permutability between the two, as proven in [9]. All these transformations corresponding to the zero multiplier preserve the class of Willmore surfaces.

The isothermic surface condition is known [8] to be preserved under constrained Willmore spectral deformation. As for Bäcklund transformation of isothermic constrained Willmore surfaces, we believe it does not necessarily preserve the isothermic condition. In contrast, the constancy of the mean curvature of a surface in 3-dimensional space-form is preserved by both constrained Willmore spectral deformation, cf. [8], and constrained Willmore Bäcklund transformation, cf. [16], for special choices of parameters, with preservation of both the space-form and the mean curvature in the latter case. However, constant mean curvature surfaces are not conformally-invariant objects, requiring that we carry a distinguished space-form. This shall be the subject of a forthcoming paper.

2.2 Constrained Willmore surfaces and perturbed harmonicity

Consider $\underline{\mathbb{C}}^{n+2} = \Sigma \times (\mathbb{R}^{n+1,1})^{\mathbb{C}}$ provided with the complex bilinear extension of the metric on $\underline{\mathbb{R}}^{n+1,1}$. In what follows, we shall make no explicit distinction between a bundle and its complexification, and move from real tensors to complex tensors by complex multilinear extension, with no need for further reference, preserving notation.

Throughout this text, we will consider the identification

$$\wedge^2\mathbb{R}^{n+1,1} \cong o(\mathbb{R}^{n+1,1})$$

of the exterior power $\wedge^2\mathbb{R}^{n+1,1}$ with the orthogonal algebra $o(\mathbb{R}^{n+1,1})$ via

$$u \wedge v(w) := (u, w)v - (v, w)u$$

for $u, v, w \in \mathbb{R}^{n+1,1}$. Given $\mu, \eta \in \Omega^1(\Sigma \times o(\mathbb{R}^{n+1,1}))$, we use $[\mu \wedge \eta]$ to denote the 2-form defined from the Lie Bracket $[\ ,\ ]$ in $o(\mathbb{R}^{n+1,1})$:

$$[\mu \wedge \eta]_{(X,Y)} = [\mu_X, \eta_Y] - [\mu_Y, \eta_X],$$

for all $X, Y \in \Gamma(T\Sigma)$. We consider the bundle $\mathrm{End}(\underline{\mathbb{R}}^{n+1,1})$, and, more generally, any bundle of morphisms, provided with the metric defined by $(\xi, \eta) := \mathrm{tr}(\eta^t \xi)$ and we shall move from a connection on $\underline{\mathbb{R}}^{n+1,1}$ to a connection on $\mathrm{End}(\underline{\mathbb{R}}^{n+1,1})$ via $\nabla\xi = \nabla \circ \xi - \xi \circ \nabla$, with preservation of notation. Note that, in the case of a metric connection ∇ on $\underline{\mathbb{R}}^{n+1,1}$, we have

$$\nabla(u \wedge v) = \nabla u \wedge v + u \wedge \nabla v,$$

for all $u, v \in \Gamma(\underline{\mathbb{R}}^{n+1,1})$.

Our theory is local and, throughout this text, with no need for further reference, restriction to a suitable non-empty open set shall be underlying.

A detailed account of elementary computations can be found in [16].

2.2.1 Real constrained Willmore surfaces

2.2.1.1 Conformal geometry of the sphere

Our study is one of surfaces in the conformal n-sphere, with $n \geq 3$, in Darboux's light-cone model [12] of the latter. For this, contemplate the light-cone $\mathcal{L}$ in the Lorentzian vector space $\mathbb{R}^{n+1,1}$ and its projectivization $\mathbb{P}(\mathcal{L})$, provided with the conformal structure defined by a metric g_σ arising from a never-zero section σ of the tautological bundle $\pi : \mathcal{L} \to \mathbb{P}(\mathcal{L})$ via

$$g_\sigma(X, Y) = (d\sigma(X), d\sigma(Y)).$$

For $v_\infty \in \mathbb{R}^{n+1,1}_\times$, set

$$S_{v_\infty} := \{v \in \mathcal{L} : (v, v_\infty) = -1\},$$

an n-dimensional submanifold of $\mathbb{R}^{n+1,1}$. Given $v \in S_{v_\infty}$,

$$T_v S_{v_\infty} = < v, v_\infty >^\perp . \tag{2.2}$$

The fact that $(v, v_\infty) \neq 0$ establishes the non-degeneracy of the subspace $< v, v_\infty >$ of $\mathbb{R}^{n+1,1}$, establishing a decomposition

$$\mathbb{R}^{n+1,1} = < v, v_\infty > \oplus T_v S_{v_\infty}. \tag{2.3}$$

In its turn, the nullity of v establishes $< v, v_\infty >$ as a 2-dimensional space with a metric with signature $(1, 1)$, showing that S_{v_∞} inherits from $\mathbb{R}^{n+1,1}$ a positive definite metric. Furthermore: for v_∞ non-null, orthoprojection onto $\langle v_\infty\rangle^\perp$ induces an isometry between S_{v_∞} and $\{v \in \langle v_\infty\rangle^\perp : (v,v) = -1/(v_\infty, v_\infty)\}$, whereas, when v_∞ is null, for any choice of $v_0 \in S_{v_\infty}$, orthoprojection onto $\langle v_0, v_\infty\rangle^\perp$ restricts to an isometry of S_{v_∞}. We conclude that S_{v_∞} inherits from $\mathbb{R}^{n+1,1}$ a positive definite metric of (constant) sectional curvature $-(v_\infty, v_\infty)$, defining a copy of the n-sphere, a copy of Euclidean n-space or two copies of hyperbolic n-space, according to the sign of (v_∞, v_∞).

By construction, the bundle projection π restricts to give a conformal diffeomorphism

$$\pi_{|S_{v_\infty}} : S_{v_\infty} \to \mathbb{P}(\mathcal{L})\backslash\mathbb{P}(\mathcal{L} \cap \langle v_\infty\rangle^\perp).$$

In particular, choosing v_∞ time-like identifies $\mathbb{P}(\mathcal{L})$ with the conformal n-sphere,

$$S^n \cong \mathbb{P}(\mathcal{L}).$$

This model linearizes the conformal geometry of the sphere. For example, k-spheres in S^n are identified with $(k+1,1)$-planes V in $\mathbb{R}^{n+1,1}$ via $V \mapsto \mathbb{P}(\mathcal{L} \cap V) \subset \mathbb{P}(\mathcal{L})$.

2.2.1.2 Surfaces in the light-cone picture and central sphere congruence

For us, a map $\Lambda : \Sigma \to \mathbb{P}(\mathcal{L})$ is the same as a null line subbundle of the trivial bundle $\underline{\mathbb{R}}^{n+1,1} = \Sigma \times \mathbb{R}^{n+1,1}$, in the natural way. From this point of view, sections of Λ are simply lifts of Λ to maps $\Sigma \to \mathbb{R}^{n+1,1}$. Given such a Λ, we define

$$\Lambda^{(1)} := \langle \sigma, d\sigma(T\Sigma)\rangle,$$

for σ a lift of Λ. For further reference, note that Λ is an immersion if and only if the bundle $\Lambda^{(1)}$ has rank 3.

Let then $\Lambda : \Sigma \to \mathbb{P}(\mathcal{L})$ be an immersion of an oriented surface Σ, which we provide with the conformal structure $\mathcal{C}_\Lambda$ induced by Λ and with J the canonical complex structure (that is, $90°$ rotation in the positive direction in the tangent spaces, a notion that is obviously invariant under conformal changes of the metric). Observe that every lift $\sigma : \Sigma \to \mathbb{R}^{n+1,1}$ of Λ is conformal: given z a holomorphic chart of Σ, $(\sigma_z, \sigma_z) = 0$ (or, equivalently, $(\sigma_{\bar z}, \sigma_{\bar z}) = 0$). Set

$$\Lambda^{1,0} := \Lambda \oplus d\sigma(T^{1,0}\Sigma), \quad \Lambda^{0,1} := \Lambda \oplus d\sigma(T^{0,1}\Sigma), \tag{2.4}$$

independently of the choice of a lift σ of Λ, defining in this way two complex rank 2 subbundles of $\Lambda^{(1)}$, complex conjugate of each other,

$$\Lambda^{0,1} = \overline{\Lambda^{1,0}}.$$

The nullity and conformality of the lifts of Λ establish the isotropy of (both) $\Lambda^{1,0}$ (and $\Lambda^{0,1}$), whilst the fact that Λ is an immersion establishes that $\Lambda^{1,0}$ and $\Lambda^{0,1}$ intersect in Λ,

$$\Lambda^{1,0} \cap \Lambda^{0,1} = \Lambda.$$

Let $S : \Sigma \to \mathcal{G} := \mathrm{Gr}_{(3,1)}(\mathbb{R}^{n+1,1})$ be the *central sphere congruence* of Λ,

$$S = \Lambda^{(1)} \oplus \langle \triangle\sigma \rangle = \langle \sigma, \sigma_z, \sigma_{\bar{z}}, \sigma_{z\bar{z}} \rangle,$$

for σ any lift of Λ, $\triangle\sigma$ the Laplacian of σ with respect to the metric g_σ and z a holomorphic chart of Σ. We use π_S and $\pi_{S^\perp}$ to denote the orthogonal projections of $\underline{\mathbb{R}}^{n+1,1}$ onto S and $S^\perp$, respectively.

Given z a holomorphic chart of Σ, let g_z denote the metric induced in Σ by z. Differentiation of $(\sigma, \sigma_z) = 0$ gives $(\sigma, \sigma_{z\bar{z}}) = -(\sigma_z, \sigma_{\bar{z}})$, which the conformality of z proves to be never-zero. In many occasions, it will be useful to consider a special choice of lift of Λ, the *normalized* lift with respect to z, the section $\sigma^z : \Sigma \to \mathcal{L}^+$ of Λ (given a choice $\mathcal{L}^+$ of one of the two connected components of $\mathcal{L}$) defined by $g_{\sigma^z} = g_z$. For further reference, note that this condition establishes, in particular, that $(\sigma^z_z, \sigma^z_{\bar{z}})$ is constant: $(\sigma^z_z, \sigma^z_{\bar{z}}) = 1/2$, and, therefore,

$$\pi_S \sigma^z_{zz} \in \Gamma((\Lambda^{(1)})^\perp \cap S) = \Gamma\Lambda. \tag{2.5}$$

Consider the decomposition of the trivial flat connection d on $\underline{\mathbb{R}}^{n+1,1}$ as

$$d = \mathcal{D} \oplus \mathcal{N}$$

for $\mathcal{D}$ the connection given by the sum of the connections ∇^S and $\nabla^{S^\perp}$ induced on S and $S^\perp$, respectively, by d. Note that $\mathcal{D}$ is a metric connection and, therefore, $\mathcal{N}$ is skew-symmetric,

$$\mathcal{N} \in \Omega^1(S \wedge S^\perp).$$

Note that, given $\xi \in \Gamma(S \wedge S^\perp)$, the transpose of $\xi|_S$ is $-\xi|_{S^\perp}$, and define a bundle isomorphism $S \wedge S^\perp \to \mathrm{Hom}(S, S^\perp)$ by $\eta \mapsto \eta|_S$. Together with the canonical identification of $S^*T\mathcal{G}$ and $\mathrm{Hom}(S, S^\perp)$, via $X \mapsto (\rho \mapsto \pi_{S^\perp}(d_X \rho))$, this gives an identification

$$S^*T\mathcal{G} \cong \mathrm{Hom}(S, S^\perp) \cong S \wedge S^\perp, \tag{2.6}$$

of bundles provided with the canonical metrics and connections (for the connection $\mathcal{D}$ on $\underline{\mathbb{R}}^{n+1,1}$), (see, for example, [10]), which we will consider throughout. Observe that, under the identification (2.6), we have

$$dS = \mathcal{N}. \tag{2.7}$$

We restrict our study to surfaces in S^n which are not contained in any subsphere of S^n. This ensures, in particular, that, given $v_\infty \in \mathbb{R}^{n+1,1}$ non-zero, $\Lambda(\Sigma) \nsubseteq \mathbb{P}(\mathcal{L} \cap \langle v_\infty \rangle^\perp)$: if v_∞ is space-like, $\mathbb{P}(\mathcal{L} \cap \langle v_\infty \rangle^\perp)$ is a hypersphere in S^n, while, in the case v_∞ is time-like or light-like, this is always necessarily the case. Hence, given σ a lift of Λ, we have $(\sigma, v_\infty) \neq 0$ and we define a local immersion

$$\sigma_\infty := (\pi_{|S_{v_\infty}})^{-1} \circ \Lambda = \frac{-1}{(\sigma, v_\infty)} \sigma : \Sigma \to S_{v_\infty},$$

of Σ into the space-form S_{v_∞}, conformally diffeomorphic to the surface Λ. The normal bundle N_∞ to σ_∞ can be identified with the normal bundle $S^\perp$ to the central sphere congruence of Λ, as bundles provided with metrics and connections:

Lemma 2.1. [8] *Let $\mathcal{H}_\infty$ denote the mean curvature vector of σ_∞. Then*

$$\xi \mapsto \xi + (\xi, \mathcal{H}_\infty)\sigma_\infty$$

defines an isomorphism

$$\mathcal{Q} : N_\infty \to S^\perp,$$

of bundles provided with a metric and a connection. Furthermore:

$$\mathcal{Q}(\mathcal{H}_\infty) = -\pi_{S^\perp}(v_\infty). \tag{2.8}$$

Proof. Let g_∞ denote the metric induced in Σ by σ_∞ and ∇^{N_∞} denote the connection induced in N_∞ by the pull-back connection by σ_∞ of the Levi-Civita connection on $(T\Sigma, g_\infty)$. According to (2.2) and (2.3), the pull-back bundle by σ_∞ of $T\Sigma$ consists of the orthogonal complement in $\underline{\mathbb{R}}^{n+1,1}$ of the non-degenerate bundle $\langle \sigma_\infty, v_\infty \rangle$,

$$\sigma_\infty^* T S_{v_\infty} = \langle \sigma_\infty, v_\infty \rangle^\perp.$$

Let π_{N_∞} denote the orthogonal projection of

$$\underline{\mathbb{R}}^{n+1,1} = d\sigma_\infty(T\Sigma) \oplus N_\infty \oplus \langle v_\infty, \sigma_\infty \rangle$$

onto N_∞. Since the metric in S_{v_∞} is the one inherited from $\mathbb{R}^{n+1,1}$, the second fundamental form Π_∞ of σ_∞ is simply given by

$$\Pi_\infty(X, Y) = \pi_{N_\infty}(d_X d_Y \sigma_\infty),$$

for $X, Y \in \Gamma(T\Sigma)$, so that, given $\xi \in \Gamma(N_\infty)$ and $(e_i)_i$ an orthonormal frame of $(T\Sigma, g_\infty)$, we have $(\xi, \sum_i d_{e_i} d_{e_i} \sigma_\infty) = 2(\xi, \mathcal{H}_\infty)$ and, therefore,

$$(\xi + (\xi, \mathcal{H}_\infty)\sigma_\infty, \sum_i d_{e_i} d_{e_i} \sigma_\infty) = 0.$$

Together with the fact that $N_\infty \subset \langle \sigma_\infty, v_\infty \rangle^\perp$, this shows that $\xi + (\xi, \mathcal{H}_\infty)\sigma_\infty$ is, in fact, a section of $S^\perp$.

Clearly, $\mathcal{Q}$ is isometric, and, therefore, injective, as N_∞ is non-degenerate. Now rank $N_\infty = n - 2 =$ rank $S^\perp$ shows that $\mathcal{Q}$ is an isometric isomorphism. Furthermore, given $\xi \in \Gamma(N_\infty)$,

$$\nabla^{S^\perp}(\mathcal{Q}(\xi)) = \pi_{S^\perp}(d\xi) + d(\xi, \mathcal{H}_\infty)\pi_{S^\perp}(\sigma_\infty) + (\xi, \mathcal{H}_\infty)\pi_{S^\perp}(d\sigma_\infty) = \pi_{S^\perp}(d\xi),$$

whilst

$$\mathcal{Q}(\nabla^{N_\infty}\xi) = \pi_{N_\infty}(d\xi) + (\pi_{N_\infty}(d\xi), \mathcal{H}_\infty)\sigma_\infty \in \Gamma(S^\perp).$$

To conclude that $\mathcal{Q}$ preserves connections, we just need to verify that

$$d\xi - \pi_{N_\infty}(d\xi) \in \Gamma(S).$$

That is immediate: $d\xi$ is still a section of $\langle \sigma_\infty, v_\infty \rangle^\perp$,

$$(d\xi, \sigma_\infty) = (d\xi, \sigma_\infty) + (\xi, d\sigma_\infty) = 0 = (d\xi, v_\infty) + (\xi, dv_\infty) = (d\xi, v_\infty);$$

and, therefore, $d\xi - \pi_{N_\infty}(d\xi)$ is the orthogonal projection of $d\xi$ onto the tangent bundle to σ_∞.

Finally, the fact that

$$(\mathcal{Q}(\xi), \pi_{S^\perp}(v_\infty)) = (\xi, v_\infty) + (\xi, \mathcal{H}_\infty)(\sigma_\infty, v_\infty) = -(\mathcal{Q}(\xi), \mathcal{Q}(\mathcal{H}_\infty)),$$

for $\xi \in N_\infty$, establishes (2.8) and completes the proof. □

2.2.1.3 The Willmore energy

Suppose, for the moment, that Σ is compact. The *Willmore energy* $\mathcal{W}(\Lambda)$ of Λ is given by[1]

$$\mathcal{W}(\Lambda) = \int_\Sigma |\Pi_0|^2 dA,$$

for Π_0 the trace-free part of the second fundamental form of Λ (calculated with respect to any representative metric on S^n and independent of that choice).

Next we present a manifestly conformally-invariant formulation of the Willmore energy. It follows the definition presented in [7], in the quaternionic setting, for the particular case of $n = 4$. The intervention of the conformal structure will restrict to the Hodge $*$-operator, which is conformally-invariant on 1-forms over a surface.

Given $\mu, \eta \in \Omega^1(S^*T\mathcal{G})$, let $(\mu \wedge \eta)$ be the 2-form defined from the metric on $S^*T\mathcal{G}$:

$$(\mu \wedge \eta)_{(X,Y)} = (\mu_X, \eta_Y) - (\mu_Y, \eta_X),$$

for all $X, Y \in \Gamma(T\Sigma)$. Note that

$$(dS \wedge *dS) = -(*dS \wedge dS) = (dS, dS)dA,$$

$(dS \wedge *dS)$ is a conformally invariant way of writing $(dS, dS)_g dA_g$, for $g \in \mathcal{C}_\Lambda$, with dA_g denoting the area element of (Σ, g) and $(\,,\,)_g$ denoting the Hilbert-Schmidt metric on $L((T\Sigma, g), S^*T\mathcal{G})$.

Theorem 2.2. [7]

$$\mathcal{W}(\Lambda) = \frac{1}{2}\int_\Sigma (dS \wedge *dS).$$

[1] In the literature, different scalings of the Willmore energy can be found. Our choice is justified by the classical scaling in the Dirichlet energy functional.

Proof. By (2.7), fixing a metric on Σ, $|dS|^2 = |\mathcal{N}|^2$. To prove the theorem, we fix $v_\infty \in \mathbb{R}^{n+1,1}$ non-zero, provide Σ with the metric induced by σ_∞ and show that $|\mathcal{N}|^2 = 2|\Pi_\infty|^2$, for Π_∞ the trace-free part of the second fundamental form of σ_∞.

Fixing a local orthonormal frame $\{X_i\}_i$ of $T\Sigma$, we have

$$|\mathcal{N}|^2 = -\sum_i \mathrm{tr}(\mathcal{N}_{X_i}\mathcal{N}_{X_i}) = 2\sum_i \mathrm{tr}(\mathcal{N}^t_{X_i}\mathcal{N}_{X_i}|_S).$$

Recall that if $(e_i)_i$ and $(\hat{e}_i)_i$ are dual basis of a vector space E provided with a metric $(\, ,)$, then, given $\mu \in \mathrm{End}(E)$, $\mathrm{tr}(\mu) = \sum_i (\mu(e_i), \hat{e}_i)$. Let $\hat{\sigma}_\infty$ be the section of S determined by the conditions $(\hat{\sigma}_\infty, \hat{\sigma}_\infty) = 0$, $(\sigma_\infty, \hat{\sigma}_\infty) = -1$ and $(\hat{\sigma}_\infty, d\sigma_\infty) = 0$. Then $(\sigma_\infty, d_{X_1}\sigma_\infty, d_{X_2}\sigma_\infty, \hat{\sigma}_\infty)$ is a frame of S with dual $(-\hat{\sigma}_\infty, d_{X_1}\sigma_\infty, d_{X_2}\sigma_\infty, -\sigma_\infty)$ and we conclude that

$$|\mathcal{N}|^2 = 2\sum_{i,j} (\mathcal{N}_{X_i}(d_{X_j}\sigma_\infty), \mathcal{N}_{X_i}(d_{X_j}\sigma_\infty)).$$

Lemma 2.1 establishes $\mathcal{N}_{X_i}(d_{X_j}\sigma_\infty) = \mathcal{Q}(\Pi_\infty(X_i, X_j))$ and completes the proof. □

2.2.1.4 Willmore surfaces and harmonicity

A conformal immersion is *Willmore* if it extremizes the Willmore functional and *constrained Willmore* if it extremizes the Willmore functional with respect to variations that infinitesimally preserve the conformal structure, that is, variations satisfying

$$\frac{d}{dt}_{|t=0}(\delta_z, \delta_z)_t = 0,$$

for the variation $(\, ,)_t$ of the induced metric and z a holomorphic chart.

Theorem 2.2 makes it clear that

$$\mathcal{W}(\Lambda) = E(S, \mathcal{C}_\Lambda),$$

the Willmore energy of Λ coincides with the Dirichlet energy of S with respect to any of the metrics in the conformal class $\mathcal{C}_\Lambda$ (although the Levi-Civita connection is not conformally-invariant, the Dirichlet energy of a mapping of a surface is preserved under conformal changes of the metric (and so is its harmonicity)). Furthermore, in a very well known result, established by Blaschke [1], for $n = 3$, and, independently, by Ejiri [13] and Rigoli [19], for general n:

Theorem 2.3. [1, 13, 19] *Λ is a Willmore surface if and only if its central sphere congruence $S : (\Sigma, \mathcal{C}_\Lambda) \to \mathcal{G}$ is a harmonic map.*

Next we present a proof of Theorem 2.3 in the light-cone picture. This is a generalization of the proof presented in [7], in the quaternionic setting, for the particular case of $n = 4$.

Proof. Given a variation Λ_t of Λ and S_t the corresponding variation of S through central sphere congruences, the Dirichlet energy $E(S_t, \mathcal{C}_t)$ of S_t with respect to the conformal structure $\mathcal{C}_t$ induced in Σ by Λ_t is given by $\frac{1}{2}\int_\Sigma (dS_t \wedge *_t dS_t)$, for $\mathrm{d}A_t$ and $*_t$ the area element and the Hodge $*$-operator of (Σ, g_t), respectively, fixing $g_t \in \mathcal{C}_t$. Hence

$$\frac{d}{dt}_{|_{t=0}} E(S_t, \mathcal{C}_t) = \frac{1}{2}\int_\Sigma ((d\dot{S} \wedge *dS) + (dS \wedge \dot{*}dS) + (dS \wedge *d\dot{S})),$$

abbreviating $\frac{d}{dt}_{|_{t=0}}$ by a dot. Let $(J_t)_t$ be the corresponding variation of J through canonical complex structures. Differentiation at $t = 0$ of $*_t dS_t = -(dS_t)J_t$ gives $\dot{*}dS = -(dS)\dot{J}$, whilst that of $J_t^2 = -I$ gives $\dot{J}J = -J\dot{J}$ and, in particular, that $\dot{J}$ intertwines the eigenspaces of J. The $\mathcal{C}_\Lambda$-conformality of S, $(d_{X \pm iJX} S, d_{X \pm iJX} S) = 0$, respectively, for $X \in \Gamma(T\Sigma)$, establishes then $(dS \wedge \dot{*}dS) = 0$ and, therefore,

$$\frac{d}{dt}_{|_{t=0}} E(S_t, \mathcal{C}_t) = \frac{d}{dt}_{|_{t=0}} E(S_t, \mathcal{C}_0).$$

It is now clear that if $S : (\Sigma, \mathcal{C}_\Lambda) \to \mathcal{G}$ is harmonic then Λ is Willmore.

Conversely, suppose Λ is Willmore, fix z a holomorphic chart of $(\Sigma, \mathcal{C}_\Lambda)$ and let us show that the tension field τ_z of $S : (\Sigma, g_z) \to \mathcal{G}$ vanishes. First of all, observe that

$$4\nabla_{\delta_z} S_{\bar{z}} = \tau_z = 4\nabla_{\delta_{\bar{z}}} S_z$$

to conclude that $\Lambda^{(1)} \subset \ker \tau_z$: by (2.5),

$$(\nabla_{\delta_z} S_{\bar{z}})\sigma_z^z = \nabla_{\delta_z}^{S^\perp}(\pi_{S^\perp}\sigma_{z\bar{z}}^z) - \pi_{S^\perp}(\nabla_{\delta_z}^S \sigma_z^z)_{\bar{z}} = 0,$$

and, similarly, $(\nabla_{\delta_{\bar{z}}} S_z)\sigma_{\bar{z}}^z = 0$, whilst $(\nabla_{\delta_z} S_{\bar{z}})\sigma^z = 0$ is immediate. It follows that $\mathrm{Im}\tau_z^t \subset \Lambda$. Fix $\Lambda_t = \langle \sigma_t \rangle$ a variation of Λ and let S_t be the corresponding variation of S through central sphere congruences. Then $\tau_z^t(\pi_{S^\perp}\dot{\sigma}) = \lambda\sigma_0$, for some $\lambda \in C^\infty(\Sigma, \mathbb{R})$, and, therefore, $\mathrm{tr}(\tau_z^t \dot{S}) = \lambda$ (note that $\dot{S}\sigma = \pi_{S^\perp}\dot{\sigma}$). On the other hand, classically,

$$0 = \frac{d}{dt}_{|_{t=0}} E(S_t, \mathcal{C}_\Lambda) = -\int_\Sigma (\dot{S}, \tau_z) dA_z = -\int_\Sigma \mathrm{tr}(\tau_z^t \dot{S}) dA_z,$$

for dA_z the area element of (Σ, g_z). Now suppose τ_z is non-zero. Then so is $\tau_z^t \in \Gamma(\mathrm{Hom}(S^\perp, S))$, so we can choose σ_t such that λ is positive, which leads to a contradiction and completes the proof. □

Having characterized Willmore surfaces by the harmonicity of the central sphere congruence, and recalling (2.7), we deduce the Willmore surface equation,

$$d^{\mathcal{D}} * \mathcal{N} = 0. \tag{2.9}$$

More generally, we have a manifestly conformally-invariant characterization of constrained Willmore surfaces in space-forms, first established in [8] and reformulated in [5] as follows:

Theorem 2.4. [5] *Λ is a constrained Willmore surface if and only if there exists a real form $q \in \Omega^1(\Lambda \wedge \Lambda^{(1)})$ with*

$$d^{\mathcal{D}} q = 0$$

such that

$$d^{\mathcal{D}} * \mathcal{N} = 2\,[q \wedge *\mathcal{N}].$$

Such a form q is said to be a [Lagrange] multiplier *for Λ and Λ is said to be a q-*constrained Willmore surface.

Proof. Calculus of variations techniques show that the variational Willmore energy relates to the variational surface by

$$\dot{\mathcal{W}} = \int_{\Sigma} ((d^{\mathcal{D}} * \mathcal{N}) \wedge \dot{\Lambda}),$$

for some non-degenerate pairing $(\wedge)$. For general variations, $\dot{\Lambda}$ can be arbitrary, establishing the Willmore surface equation (2.9), whereas infinitesimally conformal variations are characterized by the normal variational being in the image of the conformal Killing operator $\overline{\overline{\partial}}$, which, according to Weyl's Lemma, consists of $(H^0 K)^{\perp}$, for $H^0 K$ the space of holomorphic quadratic differentials. The result follows by defining a multiplier q from a quadratic differential $q^z dz^2 \in (H^0 K)^{\perp}$ via $q_{\delta_{\bar{z}}} \sigma_z = -\frac{1}{2}\, q^z \sigma$, for σ a lift of Λ and z a holomorphic chart of Σ. $\square$

Willmore surfaces are the 0-constrained Willmore surfaces. The zero multiplier is not necessarily the only multiplier for a constrained Willmore surface with no constraint on the conformal structure, though. In fact, the uniqueness of multiplier characterizes non-isothermic constrained Willmore surfaces, as we shall see below in this text.

The characterization of constrained Willmore surfaces above motivates a natural extension to surfaces that are not necessarily compact.

Next we present a useful result, which establishes, in particular, that if q is a multiplier for Λ, then $q^{1,0}$ takes values in $\Lambda \wedge \Lambda^{0,1}$.

Lemma 2.5. *Given $q \in \Omega^1(\Lambda \wedge \Lambda^{(1)})$ real,*

i) if $d^{\mathcal{D}} q = 0$ then $q^{1,0} \in \Omega^{1,0}(\Lambda \wedge \Lambda^{0,1})$ or, equivalently, $q^{0,1} \in \Omega^{0,1}(\Lambda \wedge \Lambda^{1,0})$;

ii) $d^{\mathcal{D}} q = 0$ if and only if $d^{\mathcal{D}} q^{1,0} = 0$, or, equivalently, $d^{\mathcal{D}} q^{0,1} = 0$;

iii) $d^{\mathcal{D}} q = 0$ if and only if $d^{\mathcal{D}} * q = 0$.

Proof. Fix z a holomorphic chart of Σ. First of all, observe that a section ξ of $\Lambda \wedge \Lambda^{(1)}$ is a section of $\Lambda \wedge \Lambda^{1,0}$ if and only if $\xi(\sigma_{\bar{z}}^{z}) = 0$. Suppose $d^{\mathcal{D}} q = 0$. Then, in particular, $d^{\mathcal{D}} q\,(\delta_z, \delta_{\bar{z}})\,\sigma^z = 0$, or, equivalently,

$$\mathcal{D}_{\delta_z}(q_{\delta_{\bar{z}}} \sigma^z) - q_{\delta_{\bar{z}}}(\mathcal{D}_{\delta_z} \sigma^z) - \mathcal{D}_{\delta_{\bar{z}}}(q_{\delta_z} \sigma^z) + q_{\delta_z}(\mathcal{D}_{\delta_{\bar{z}}} \sigma^z) = 0,$$

establishing

$$q_{\delta_z}\sigma^z_{\bar z} = q_{\delta_{\bar z}}\sigma^z_{z}. \tag{2.10}$$

In its turn, $d^{\mathcal{D}}q\,(\delta_z, \delta_{\bar z})\,\sigma^z_z = 0$ implies

$$\mathcal{D}_{\delta_z}(q_{\delta_{\bar z}}\sigma^z_z) - \mathcal{D}_{\delta_{\bar z}}(q_{\delta_z}\sigma^z_z) + q_{\delta_z}\sigma^z_{z\bar z} = 0,$$

by (2.5). On the other hand, the skew-symmetry of q establishes $(q\sigma^z_{z\bar z}, \sigma^z_{z\bar z}) = 0$ and, therefore,

$$q\sigma^z_{z\bar z} = \mu\sigma_z + \eta\sigma^z_{\bar z}, \tag{2.11}$$

for some $\mu, \eta \in \Omega^1(\underline{\mathbb{C}})$. Hence

$$\mathcal{D}_{\delta_z}(q_{\delta_{\bar z}}\sigma^z_z) + \mu_{\delta_z}\sigma^z_z = \mathcal{D}_{\delta_{\bar z}}(q_{\delta_z}\sigma^z_z) - \eta_{\delta_z}\sigma^z_{\bar z}.$$

It is obvious that a section of $\Lambda \wedge \Lambda^{(1)}$ transforms sections of $\Lambda^{(1)}$ into sections of Λ, so that, in particular, both $q_{\delta_{\bar z}}\sigma^z_z$ and $q_{\delta_z}\sigma^z_z$ are sections of Λ. We conclude that $\mathcal{D}_{\delta_z}(q_{\delta_{\bar z}}\sigma^z_z) + \mu_{\delta_z}\sigma^z_z$ is a section of $\Lambda^{1,0} \cap \Lambda^{0,1} = \Lambda$. Write $q_{\delta_{\bar z}}\sigma^z_z = \lambda\sigma^z$, with $\lambda \in \Gamma(\underline{\mathbb{C}})$. Then

$$\lambda_z\sigma^z + (\lambda + \mu_{\delta_z})\sigma^z_z = \gamma\sigma^z,$$

for some $\gamma \in \Gamma(\underline{\mathbb{C}})$. In particular, $\lambda = -\mu_{\delta_z}$. Equation (2.11) establishes, on the other hand,

$$q = -2\mu\,\sigma^z \wedge \sigma^z_z - 2\eta\,\sigma^z \wedge \sigma^z_{\bar z}$$

and, in particular, $q_{\delta_z}\sigma^z_{\bar z} = \mu_{\delta_z}\sigma^z$. Equation (2.10) completes the proof of *i*).

Next we prove *ii*). By (2.5),

$$\mathcal{D}^{1,0}\Gamma(\Lambda^{1,0}) \subset \Omega^{1,0}(\Lambda^{1,0})$$

or, equivalently,

$$\mathcal{D}^{0,1}\Gamma(\Lambda^{0,1}) \subset \Omega^{0,1}(\Lambda^{0,1})$$

and, therefore, following *i*), $d^{\mathcal{D}}q^{1,0} \in \Omega^2(\Lambda \wedge \Lambda^{0,1})$ and $d^{\mathcal{D}}q^{0,1} \in \Omega^2(\Lambda \wedge \Lambda^{1,0})$. Hence $d^{\mathcal{D}}q = 0$ forces $d^{\mathcal{D}}q^{1,0}$ and $d^{\mathcal{D}}q^{0,1}$ to vanish separately. The reality of q completes the proof of *ii*).

As for *iii*), it is immediate from *ii*). □

2.2.1.5 Constrained Willmore surfaces: a zero-curvature characterization

For a map into a Grassmannian, harmonicity amounts to the flatness of a family of connections, according to Uhlenbeck [22]. With the characterization of Willmore surfaces by the harmonicity of the central sphere congruence, a zero-curvature characterization of Wilmore surfaces follows. More generally, the constrained Willmore surface equations amount to the flatness of a certain[2] family of connections:

---

[2]The associated family of flat connections presented in [9] corresponds to a different choice of orientation in Σ.

Theorem 2.6. [5] *Λ is a constrained Willmore surface if and only if there exists a real form $q \in \Omega^1(\Lambda \wedge \Lambda^{(1)})$ such that*

$$d_q^\lambda := \mathcal{D} + \lambda^{-1}\mathcal{N}^{1,0} + \lambda\mathcal{N}^{0,1} + (\lambda^{-2} - 1)q^{1,0} + (\lambda^2 - 1)q^{0,1}$$

is flat for all $\lambda \in S^1$.

Before proceeding to the proof of the theorem, and for further reference, observe that, given $a, b \in \mathbb{R}^{n+1,1}$ and $T \in o(\mathbb{R}^{n+1,1})$,

$$[T, a \wedge b] = (Ta) \wedge b + a \wedge (Tb),$$

to conclude that

$$[\Lambda \wedge \Lambda^{(1)}, \Lambda \wedge \Lambda^{(1)}] \subset \Lambda \wedge \Lambda = \{0\}. \tag{2.12}$$

Now we proceed to the proof of the theorem.

Proof. According to the decomposition

$$o(\underline{\mathbb{R}}^{n+1,1}) = (\wedge^2 S \oplus \wedge^2 S^\perp) \oplus S \wedge S^\perp, \tag{2.13}$$

the flatness of d, characterized by

$$0 = R^{\mathcal{D}} + d^{\mathcal{D}}\mathcal{N} + \frac{1}{2}[\mathcal{N} \wedge \mathcal{N}],$$

encodes two structure equations, namely,

$$R^{\mathcal{D}} + \frac{1}{2}[\mathcal{N} \wedge \mathcal{N}] = 0 \tag{2.14}$$

and

$$d^{\mathcal{D}}\mathcal{N} = 0. \tag{2.15}$$

Now suppose $q \in \Omega^1(\Lambda \wedge \Lambda^{(1)})$ is a real form. Given $\lambda \in S^1$, set

$$A^\lambda = d_q^\lambda - \mathcal{D} \in \Omega^1(\mathrm{End}(\underline{R}^{n+1,1})).$$

The curvature tensor of d_q^λ is given by

$$R^{d_q^\lambda} = R^{\mathcal{D}} + d^{\mathcal{D}}A^\lambda + \frac{1}{2}[A^\lambda \wedge A^\lambda].$$

Since there are no non-zero $(2,0)$- or $(0,2)$-forms over a surface, we have

$$\begin{aligned}\frac{1}{2}[A^\lambda \wedge A^\lambda] =& [\mathcal{N}^{1,0} \wedge \mathcal{N}^{0,1}] + (\lambda^{-1} - \lambda)\big([q^{1,0} \wedge \mathcal{N}^{0,1}] \\ &- [q^{0,1} \wedge \mathcal{N}^{1,0}]\big) + (2 - \lambda^{-2} - \lambda^2)[q^{1,0} \wedge q^{0,1}]\end{aligned}$$

and equation (2.14) establishes then

$$R^{d_q^\lambda} = d^{\mathcal{D}}A^\lambda + (\lambda^{-1} - \lambda)\big([q^{1,0} \wedge \mathcal{N}^{0,1}]$$

$$- [q^{0,1} \wedge \mathcal{N}^{1,0}]) + \frac{1}{2}(2 - \lambda^{-2} - \lambda^2)[q \wedge q].$$

But, according to (2.12), $[q \wedge q] = 0$. Hence

$$R^{d_q^\lambda} = d^{\mathcal{D}} A^\lambda + (\lambda^{-1} - \lambda)\big([q^{1,0} \wedge \mathcal{N}^{0,1}] - [q^{0,1} \wedge \mathcal{N}^{1,0}]\big).$$

In its turn, equation (2.15) gives

$$d^{\mathcal{D}} \mathcal{N}^{1,0} = \frac{i}{2} d^{\mathcal{D}} * \mathcal{N} = -d^{\mathcal{D}} \mathcal{N}^{0,1}.$$

We conclude that

$$R^{d_q^\lambda} = \frac{\lambda^{-1} - \lambda}{2} i\,(d^{\mathcal{D}} * \mathcal{N} - 2[q \wedge *\mathcal{N}]) + (\lambda^{-2} - 1)\, d^{\mathcal{D}} q^{1,0} + (\lambda^2 - 1)\, d^{\mathcal{D}} q^{0,1}.$$

Yet again, according to the decomposition (2.13), it follows that $R^{d_q^\lambda} = 0$ if and only if both

$$\frac{\lambda^{-1} - \lambda}{2} i\,(d^{\mathcal{D}} * \mathcal{N} - 2[q \wedge *\mathcal{N}]) = 0 \tag{2.16}$$

and

$$(\lambda^{-2} - 1)\, d^{\mathcal{D}} q^{1,0} + (\lambda^2 - 1)\, d^{\mathcal{D}} q^{0,1} = 0 \tag{2.17}$$

hold. Organizing equations (2.16) and (2.17) by powers of λ completes the proof. □

2.2.1.6 Constrained Willmore surfaces and the isothermic surface condition

Isothermic surfaces are classically defined by the existence of conformal curvature line coordinates. Equation (2.1) makes it clear that, although the second fundamental form is not conformally invariant, conformal curvature line coordinates are preserved under conformal changes of the metric and, therefore, so is the isothermic surface condition. The next result presents a manifestly conformally-invariant formulation of the isothermic surface condition, established in [6] and discussed also in [11].

Lemma 2.7. [6, 11] *Λ is isothermic if and only if there exists a non-zero closed real 1-form $\eta \in \Omega^1(\Lambda \wedge \Lambda^{(1)})$. Under these conditions, we may say that (Λ, η) is an isothermic surface. The form η is defined up to a real constant scale.*

Remark 1. *According to the decomposition (2.13), given $\eta \in \Omega^1(\Lambda \wedge \Lambda^{(1)})$, $d\eta = d^{\mathcal{D}}\eta + [\mathcal{N} \wedge \eta]$ vanishes if and only if $d^{\mathcal{D}}\eta = 0 = [\mathcal{N} \wedge \eta]$.*

Proposition 2.8. [9]*A constrained Willmore surface has a unique multiplier if and only if it is not an isothermic surface. Furthermore:*

i) if $q_1 \neq q_2$ are multipliers for Λ, then $(\Lambda, *(q_1 - q_2))$ is isothermic;

ii) if (Λ, η) is an isothermic q-constrained Willmore surface, then the set of multipliers to Λ is the affine space $q + \langle *\eta \rangle_{\mathbb{R}}$.

Proof. It is immediate, noting that $[\mathcal{N} \wedge \eta] = [*\eta \wedge *\mathcal{N}]$ and recalling Lemma 2.5 - *iii*). □

A classical result by Thomsen [21] characterizes isothermic Willmore surfaces in 3-space as minimal surfaces in some 3-dimensional space-form. Constant mean curvature surfaces in 3-dimensional space-forms are examples of isothermic constrained Willmore surfaces, as proven by J. Richter [18]. However, isothermic constrained Willmore surfaces in 3-space are not necessarily constant mean curvature surfaces in some space-form, as established by an example, presented in [2], of a constrained Willmore cylinder that does not have constant mean curvature in any space-form.

2.2.2 Complexified constrained Willmore surfaces

The transformations of a constrained Willmore surface Λ we present below in this work are, in particular, pairs $((\Lambda^{1,0})^*, (\Lambda^{0,1})^*)$ of transformations $(\Lambda^{1,0})^*$ and $(\Lambda^{0,1})^*$ of $\Lambda^{1,0}$ and $\Lambda^{0,1}$, respectively. The fact that $\Lambda^{1,0}$ and $\Lambda^{0,1}$ intersect in a rank 1 bundle will ensure that $(\Lambda^{1,0})^*$ and $(\Lambda^{0,1})^*$ have the same property. The isotropy of $\Lambda^{1,0}$ and $\Lambda^{0,1}$ will ensure that of $(\Lambda^{1,0})^*$ and $(\Lambda^{0,1})^*$ and, therefore, that of their intersection. The reality of the bundle $\Lambda^{1,0} \cap \Lambda^{0,1}$ is preserved by the spectral deformation, but it is not clear that the same is necessarily true for Bäcklund transformation. This motivates the definition of *complexified surface*.

Fix a conformal structure $\mathcal{C}$ on Σ and consider the corresponding complex structure on Σ. Let $\hat{d}$ be a flat metric connection on $\underline{\mathbb{C}}^{n+2}$ and d denote the trivial flat connection. In what follows, omitting the reference to some specific connection shall be understood as an implicit reference to d.

Definition 2.9. *We define a* complexified $\hat{d}$-surface *to be a pair* $(\Lambda^{1,0}, \Lambda^{0,1})$ *of isotropic rank* 2 *subbundles of* $\underline{\mathbb{C}}^{n+2}$ *intersecting in a rank* 1 *bundle*

$$\Lambda := \Lambda^{1,0} \cap \Lambda^{0,1}$$

such that

$$\hat{d}^{1,0}\Gamma\Lambda \subset \Omega^{1,0}\Lambda^{1,0}, \quad \hat{d}^{0,1}\Gamma\Lambda \subset \Omega^{0,1}\Lambda^{0,1}.$$

Obviously, given Λ a (real) surface in $\mathbb{P}(\mathcal{L})$, $(\Lambda^{1,0}, \Lambda^{0,1})$ is a complexified surface with respect to $\mathcal{C}_\Lambda$. Henceforth, we drop the term "complexified" and use *real surface* when referring explicitly to a complexified surface $(\Lambda^{1,0}, \Lambda^{0,1})$ defining a real surface Λ. Observe that $(\Lambda^{1,0}, \Lambda^{0,1})$ is a real surface if and only if Λ is a real bundle (recall that Λ is an immersion if and only if the bundle $\Lambda^{1,0} + \Lambda^{0,1}$ has rank 3).

Observe, on the other hand, that, in the particular case of a real surface $(\Lambda^{1,0}, \Lambda^{0,1})$, the notation $\Lambda^{1,0}$ and $\Lambda^{0,1}$ is consistent with (2.4). Indeed, the $\mathcal{C}$-isotropy of $\Lambda^{1,0}$ characterizes the $\mathcal{C}$-conformality of the lifts of Λ, or equivalently, the fact that $\mathcal{C} = \mathcal{C}_\Lambda$.

2.2.2.1 Constrained Willmore surfaces and perturbed harmonic bundles

Theorem 2.6 motivates the definition of *perturbed harmonicity* for a bundle, which we present next and which will apply to the central sphere congruence to provide a characterization of constrained Willmore surfaces. In the particular case of a bundle of $(3,1)$-planes in $\mathbb{R}^{n+1,1}$, our notion of perturbed harmonicity coincides with the notion of 2-*perturbed harmonicity* introduced in [9].

Given V a non-degenerate subbundle of $\underline{\mathbb{C}}^{n+2}$, consider the decomposition $\hat{d} = \mathcal{D}_V^{\hat{d}} + \mathcal{N}_V^{\hat{d}}$ for $\mathcal{D}_V^{\hat{d}}$ the metric connection on $\underline{\mathbb{C}}^{n+2}$ given by the sum of the connections induced on V and $V^{\perp}$ by $\hat{d}$.

Definition 2.10. *A non-degenerate rank 4 subbundle V of $\underline{\mathbb{C}}^{n+2}$ is said to be a central sphere congruence of a $\hat{d}$-surface $(\Lambda^{1,0}, \Lambda^{0,1})$ if*

$$\Lambda^{1,0} + \Lambda^{0,1} \subset V$$

and

$$(\mathcal{N}_V^{\hat{d}})^{1,0}\Lambda^{0,1} = 0 = (\mathcal{N}_V^{\hat{d}})^{0,1}\Lambda^{1,0}.$$

Remark 2. *Let $(\Lambda^{1,0}, \Lambda^{0,1})$ be a surface, $\sigma \neq 0$ be a section of Λ and z be a holomorphic chart of Σ. Note that $\Lambda^{1,0} + \Lambda^{0,1} \subset V$ establishes*

$$(\mathcal{N}_V)_{|\Lambda} = 0$$

and then $\mathcal{N}_V^{1,0}\Lambda^{0,1} = 0 = \mathcal{N}_V^{0,1}\Lambda^{1,0}$ reads $\sigma_{z\bar{z}} \in \Gamma V$. Hence, generically (if $\sigma \wedge \sigma_z \neq 0 \neq \sigma \wedge \sigma_{\bar{z}}$), $\Lambda^{1,0}$, $\Lambda^{0,1}$ and V are all determined by Λ: $\Lambda^{1,0} = \langle \sigma, \sigma_z \rangle$, $\Lambda^{0,1} = \langle \sigma, \sigma_{\bar{z}} \rangle$ and

$$V = \langle \sigma, \sigma_z, \sigma_{\bar{z}}, \sigma_{z\bar{z}} \rangle.$$

In particular, the complexification of the central sphere congruence of a real surface Λ is the unique central sphere congruence of the corresponding surface $(\Lambda^{1,0}, \Lambda^{0,1})$.

For further reference:

Lemma 2.11. *Suppose V is a central sphere congruence of a $\hat{d}$-surface $(\Lambda^{1,0}, \Lambda^{0,1})$. Then*

$$(\mathcal{D}_V^{\hat{d}})^{1,0}\Gamma\Lambda^{1,0} \subset \Omega^{1,0}\Lambda^{1,0}, \quad (\mathcal{D}_V^{\hat{d}})^{0,1}\Gamma\Lambda^{0,1} \subset \Omega^{0,1}\Lambda^{0,1}.$$

Proof. First of all, observe that, as $\operatorname{rank}\Lambda^{1,0} = \frac{1}{2}\operatorname{rank} V = \operatorname{rank}\Lambda^{0,1}$, the isotropy of both $\Lambda^{1,0}$ and $\Lambda^{0,1}$ establishes their maximal isotropy in V. Write $\Lambda^{1,0} = \langle \sigma, \tau \rangle$, with $\sigma \in \Gamma\Lambda$. Since

$$(\mathcal{D}_V^{\hat{d}})^{1,0}\sigma = \pi_V \circ \hat{d}^{1,0} \circ \pi_V \sigma \in \Omega^{1,0}\Lambda^{1,0},$$

the fact that $\mathcal{D}_V^{\hat{d}}$ is a metric connection, together with the isotropy of $\Lambda^{1,0}$, shows that

$$((\mathcal{D}_V^{\hat{d}})^{1,0}\tau, \sigma) = -(\tau, (\mathcal{D}_V^{\hat{d}})^{1,0}\sigma) = 0,$$

whereas

$$((\mathcal{D}_V^{\hat{d}})^{1,0}\tau,\tau)=\frac{1}{2}\,d^{1,0}(\tau,\tau)=0.$$

We conclude that $(\mathcal{D}_V^{\hat{d}})^{1,0}\tau\perp\Lambda^{1,0}$ and, therefore, that $(\mathcal{D}_V^{\hat{d}})^{1,0}\tau$ takes values in $\Lambda^{1,0}$. A similar argument establishes $(\mathcal{D}_V^{\hat{d}})^{0,1}\Gamma\Lambda^{0,1}\subset\Omega^{0,1}\Lambda^{0,1}$. □

Definition 2.12. *A non-degenerate bundle $V\subset\underline{\mathbb{C}}^{n+2}$ is said to be $\hat{d}$-*perturbed harmonic *if there exists a 1-form q with values in $\wedge^2V\oplus\wedge^2V^\perp$ such that, for each $\lambda\in\mathbb{C}\backslash\{0\}$, the metric connection*

$$\hat{d}_V^{\lambda,q}:=\mathcal{D}_V^{\hat{d}}+\lambda^{-1}(\mathcal{N}_V^{\hat{d}})^{1,0}+\lambda(\mathcal{N}_V^{\hat{d}})^{0,1}+(\lambda^{-2}-1)q^{1,0}+(\lambda^2-1)q^{0,1},$$

*on $\underline{\mathbb{C}}^{n+2}$, is flat. In this case, we say that V is $(q,\hat{d})$-*perturbed harmonic *or, in the case $q=0$, $\hat{d}$-*harmonic. *In the particular case of $\hat{d}=d$, V a real bundle and q a real form, we say that V is a* real q-perturbed harmonic bundle.

Definition 2.13. *A $\hat{d}$-surface $(\Lambda^{1,0},\Lambda^{0,1})$ is said to be $\hat{d}$-*constrained Willmore *if it admits a $(q,\hat{d})$-perturbed harmonic central sphere congruence with*

$$q^{1,0}\in\Omega^{1,0}(\wedge^2\Lambda^{0,1}),\quad q^{0,1}\in\Omega^{0,1}(\wedge^2\Lambda^{1,0}).\tag{2.18}$$

*If $q=0$, we say that $(\Lambda^{1,0},\Lambda^{0,1})$ is $\hat{d}$-*Willmore. *In the particular case of $\hat{d}=d$, $(\Lambda^{1,0},\Lambda^{0,1})$ a real surface and q a real form, we say that $(\Lambda^{1,0},\Lambda^{0,1})$ is a* real q-constrained Willmore surface.

From Theorem 2.6 and Lemma 2.5, it follows that the real constrained Willmore surface condition is preserved under the correspondence

$$(\Lambda^{1,0},\Lambda^{0,1})\longleftrightarrow\Lambda^{1,0}\cap\Lambda^{0,1}$$

for real surfaces $(\Lambda^{1,0},\Lambda^{0,1})$, with preservation of multipliers:

Theorem 2.14. [9]*Suppose $(\Lambda^{1,0},\Lambda^{0,1})$ is a real surface. Then Λ is constrained Willmore if and only if S is q-perturbed harmonic, for some real 1-form q with $q^{1,0}\in\Omega^{1,0}(\wedge^2\Lambda^{0,1})$.*

2.3 Transformations of perturbed harmonic bundles and constrained Willmore surfaces

Fix a conformal structure $\mathcal{C}$ on Σ and consider the corresponding complex structure on Σ. Let Ad denote the adjoint representation of the orthogonal group on the orthogonal algebra. Note that, given $T\in O(\mathbb{R}^{n+1,1})$ and $u,v\in\mathbb{R}^{n+1,1}$,

$$\mathrm{Ad}_T(u\wedge v)=Tu\wedge Tv.$$

Let V be a non-degenerate subbundle of $\underline{\mathbb{C}}^{n+2}$ and π_V and $\pi_{V^\perp}$ denote the orthogonal projections of $\underline{\mathbb{C}}^{n+2}$ onto V and $V^\perp$, respectively, and ρ denote reflection across V,

$$\rho = \pi_V - \pi_{V^\perp}.$$

Let $(\Lambda^{1,0}, \Lambda^{0,1})$ be a surface admitting V as a central sphere congruence. As usual, we write Λ for $\Lambda^{1,0} \cap \Lambda^{0,1}$. Suppose V is q-perturbed harmonic for some $q \in \Omega^1(\wedge^2 V \oplus \wedge^2 V^\perp)$ satisfying conditions (2.18).

2.3.1 Spectral deformation

For each $\lambda \in \mathbb{C}\backslash\{0\}$, the flatness of the metric connection $d_V^{\lambda,q}$ on $\underline{\mathbb{C}}^{n+2}$ establishes the existence of an isometry

$$\phi_V^{\lambda,q} : (\underline{\mathbb{C}}^{n+2}, d_V^{\lambda,q}) \to (\underline{\mathbb{C}}^{n+2}, d)$$

of bundles, preserving connections, defined on a simply connected component of Σ and unique up to a Möbius transformation.

Lemma 2.15. *Let $\hat{d}$ be a flat metric connection on $\underline{\mathbb{C}}^{n+2}$ and*

$$\phi : (\underline{\mathbb{C}}^{n+2}, \hat{d}) \to (\underline{\mathbb{C}}^{n+2}, d)$$

be an isometry of bundles, preserving connections. Then V is $(q, \hat{d})$-perturbed harmonic, for some q, if and only if ϕV is $\mathrm{Ad}_\phi q$-perturbed harmonic.

Proof. It is immediate from the fact that

$$\mathcal{D}_{\phi V} = \phi \circ \mathcal{D}_V^{\hat{d}} \circ \phi^{-1}, \quad \mathcal{N}_{\phi V} = \phi \mathcal{N}_V^{\hat{d}} \phi^{-1} \tag{2.19}$$

and, therefore,

$$d_{\phi V}^{\lambda,q} = \phi \circ \hat{d}_V^{\,\lambda,\mathrm{Ad}_{\phi^{-1}} q} \circ \phi^{-1}.$$

□

Set

$$q_\lambda := \lambda^{-2} q^{1,0} + \lambda^2 q^{0,1},$$

for $\lambda \in \mathbb{C}\backslash\{0\}$. The fact that q takes values in $\wedge^2 V \oplus \wedge^2 V^\perp$ establishes, in particular,

$$\mathcal{D}_V^{d_V^{\lambda,q}} = \mathcal{D}_V + (\lambda^{-2} - 1) q^{1,0} + (\lambda^{0,1} - 1) q^{0,1},$$

whereas

$$\mathcal{N}_V^{d_V^{\lambda,q}} = \lambda^{-1} \mathcal{N}_V^{1,0} + \lambda \mathcal{N}_V^{0,1},$$

and, therefore,

$$(d_V^{\lambda,q})_V^{\mu,q_\lambda} = d_V^{\lambda\mu,q},$$

for all $\lambda, \mu \in \mathbb{C}\backslash\{0\}$. From the flatness of $d_V^{\lambda,q}$, for all $\lambda \in \mathbb{C}\backslash\{0\}$, we conclude that of $(d_V^{\lambda,q})_V^{\mu,q_\lambda}$, for all $\lambda, \mu \in \mathbb{C}\backslash\{0\}$ and, therefore, that V is $d_V^{\lambda,q}$-perturbed harmonic, for all $\lambda \in \backslash\mathbb{C}\{0\}$. We define then a spectral deformation of V into new perturbed harmonic bundles by setting, for each λ in $\mathbb{C}\{0\}$,

$$V_q^\lambda := \phi_V^{\lambda,q} V.$$

Theorem 2.16. [9] *V_q^λ is $\mathrm{Ad}_{\phi_q^\lambda} q_\lambda$-perturbed harmonic, for each $\lambda \in \mathbb{C}\{0\}$.*

A deformation on the level of constrained Willmore surfaces follows:

Theorem 2.17. [9] *For each $\lambda \in \mathbb{C}\backslash\{0\}$, $(\phi_V^{\lambda,q}\,\Lambda^{1,0}, \phi_V^{\lambda,q}\,\Lambda^{0,1})$ is a $\mathrm{Ad}_{\phi_q^\lambda}(q_\lambda)$-constrained Willmore surface, admitting V_q^λ as a central sphere congruence. Furthermore, if $(\Lambda^{1,0}, \Lambda^{0,1})$ is a real constrained Willmore surface, then so is*

$$\Lambda_q^\lambda := \phi_V^{\lambda,q}\,\Lambda,$$

for all $\lambda \in S^1$.

Proof. By (2.18), together with the isotropy of $\Lambda^{i,j}$, for $i \neq j \in \{0,1\}$, we have $q\Lambda = 0$. On the other hand, the centrality of V with respect to $(\Lambda^{1,0}, \Lambda^{0,1})$ gives $\mathcal{N}_V \Lambda = \pi_{V^\perp} \circ d\Lambda = 0$. Hence

$$(d_V^{\lambda,q})_{|\Gamma\Lambda} = d_{|\Gamma\Lambda} \tag{2.20}$$

and we conclude that $(\phi_V^{\lambda,q}\,\Lambda^{1,0}, \phi_V^{\lambda,q}\,\Lambda^{0,1})$ is still a surface. Suppose, furthermore, that $(\Lambda^{1,0}, \Lambda^{0,1})$ is a real q-constrained Willmore surface. Given $\lambda \in S^1$, d_q^λ is real, so that we can choose $\phi_V^{\lambda,q}$ to be real, in which case

$$\overline{\phi_V^{\lambda,q}\Lambda} = \phi_V^{\lambda,q}\Lambda,$$

Λ_q^λ is a real surface. It is obvious, on the other hand, that as V is a central sphere congruence of $(\Lambda^{1,0}, \Lambda^{0,1})$, $\phi_V^{\lambda,q}\,V$ is a central sphere congruence of $(\phi_V^{\lambda,q}\,\Lambda^{1,0}, \phi_V^{\lambda,q}\,\Lambda^{0,1})$. Theorem 2.16 completes the proof. □

Note that spectral deformation corresponding to the zero multiplier preserves the class of Willmore surfaces.

This spectral deformation of real constrained Willmore surfaces coincides, up to reparametrization, with the one presented in [8], in terms of the *Schwarzian derivative* and the *Hopf differential* (see [16, Section 6.4.1]).

An alternative perspective on this spectral deformation of perturbed harmonic bundles and constrained Willmore surfaces is that of a change of flat connection on $\underline{\mathbb{C}}^{n+2}$: if $V \subset (\underline{\mathbb{C}}^{n+2}, d)$ is perturbed harmonic, then so is $V \subset (\underline{\mathbb{C}}^{n+2}, d_V^{\lambda,q})$, as well as, if $(\Lambda^{1,0}, \Lambda^{0,1})$ is constrained Willmore [with respect to d] then $(\Lambda^{1,0}, \Lambda^{0,1})$ is still constrained Willmore with respect to $d_V^{\lambda,q}$, for all $\lambda \in \mathbb{C}\backslash\{0\}$. In the real case, this is the interpretation of loop group theory in [5].

2.3.2 Dressing action

We use a version of the dressing action theory of Terng and Uhlenbeck [20] to build transformations of V into new perturbed harmonic bundles and thereafter transformations of $(\Lambda^{1,0}, \Lambda^{0,1})$ into new constrained Willmore surfaces. For that, we give conditions on a dressing $r(\lambda) \in \Gamma(O(\underline{\mathbb{C}}^{n+2}))$ such that the gauging $r(\lambda) \circ d_V^{\lambda,q} \circ r(\lambda)^{-1}$ of $d_V^{\lambda,q}$ by $r(\lambda)$ establishes the perturbed harmonicity of some bundle $\hat{V}$ from the perturbed harmonicity of V.

The $\mathcal{D}_V$-parallelness of V and $V^\perp$, together with the fact that $\mathcal{N}_V$ intertwines V and $V^\perp$, whereas q preserves them, makes clear that

$$d_V^{-\lambda,q} = \rho \circ d_V^{\lambda,q} \circ \rho^{-1}, \tag{2.21}$$

for $\lambda \in \mathbb{C}\backslash\{0\}$. Suppose we have $r(\lambda) \in \Gamma(O(\underline{\mathbb{C}}^{n+2}))$ such that $\lambda \mapsto r(\lambda)$ is rational in λ, r is holomorphic and invertible at $\lambda = 0$ and $\lambda = \infty$ and twisted in the sense that

$$\rho\, r(\lambda)\, \rho^{-1} = r(-\lambda), \tag{2.22}$$

for $\lambda \in \mathrm{dom}(r)$. In particular, it follows that both $r(0)$ and $r(\infty)$ commute with ρ, and, therefore, that

$$r(0)_{|_V}, r(\infty)_{|_V} \in \Gamma(O(V)).$$

Define $\hat{q} \in \Omega^1(\wedge^2 V \oplus \wedge^2 V^\perp)$ by setting

$$\hat{q}^{1,0} := \mathrm{Ad}_{r(0)} q^{1,0}, \quad \hat{q}^{\,0,1} := \mathrm{Ad}_{r(\infty)} q^{0,1}.$$

Define a new family of metric connections on $\underline{\mathbb{C}}^{n+2}$ by setting

$$\hat{d}_V^{\lambda,\hat{q}} := r(\lambda) \circ d_V^{\lambda,q} \circ r(\lambda)^{-1}.$$

Suppose that there exists a holomorphic extension of $\lambda \mapsto \hat{d}_V^{\lambda,\hat{q}}$ to $\lambda \in \mathbb{C}\backslash\{0\}$ through metric connections on $\underline{\mathbb{C}}^{n+2}$. We shall see later how to construct such $r = r(\lambda)$, but assume, for the moment, that we have got one. In that case, as we, crucially, verify next, the notation $\hat{d}_V^{\lambda,\hat{q}}$ is not merely formal:

Proposition 2.18. [9]

$$\hat{d}_V^{\lambda,\hat{q}} = \mathcal{D}_V^{\hat{d}} + \lambda^{-1}(\mathcal{N}_V^{\hat{d}})^{1,0} + \lambda(\mathcal{N}_V^{\hat{d}})^{0,1} + (\lambda^{-2} - 1)\hat{q}^{1,0} + (\lambda^2 - 1)\hat{q}^{0,1},$$

for the flat metric connection $\hat{d} := \hat{d}_V^{1,\hat{q}} = \lim_{\lambda \to 1} r(\lambda) \circ d_V^{\lambda,q} \circ r(\lambda)^{-1}$ *and* $\lambda \in \mathbb{C}\backslash\{0\}$.

Proof. The fact that r is holomorphic and invertible at $\lambda = 0$ and that $(d_V^{\lambda,q})^{0,1} = \mathcal{D}_V^{0,1} + \lambda\mathcal{N}_V^{0,1} + (\lambda^2 - 1)q^{0,1}$ is holomorphic on $\mathbb{C}$ establishes that the connection

$$(\hat{d}_V^{\lambda,\hat{q}})^{0,1} = r(\lambda) \circ (d_V^{\lambda,q})^{0,1} \circ r(\lambda)^{-1},$$

which admits a holomorphic extension to $\lambda \in \mathbb{C}\backslash\{0\}$, admits, furthermore, a holomorphic extension to $\lambda \in \mathbb{C}$. Thus, locally,

$$(d_V^{\hat{\lambda},\hat{q}})^{0,1} = A_0^{0,1} + \sum_{i\geq 1} \lambda^i A_i^{0,1},$$

with A_0 connection and $A_i \in \Omega^1(o(\underline{\mathbb{C}}^{n+2}))$, for all i. Considering then limits of

$$\lambda^{-2} A_0^{0,1} + \sum_{i\geq 1} \lambda^{i-2} A_i^{0,1} = r(\lambda) \circ (\lambda^{-2}\mathcal{D}_V^{0,1} + \lambda^{-1}\mathcal{N}_V^{0,1} + (1-\lambda^{-2})q^{0,1}) \circ r(\lambda)^{-1},$$

when λ goes to infinity, we get

$$A_2^{0,1} + \lim_{\lambda\to\infty} \sum_{i\geq 3} \lambda^{i-2} A_i^{0,1} = \mathrm{Ad}_{r(\infty)}\, q^{0,1},$$

which shows that $A_i^{0,1} = 0$, for all $i \geq 3$, and that $A_2^{0,1} = \hat{q}^{0,1}$. Considering now limits of

$$A_0^{0,1} + \lambda A_1^{0,1} + \lambda^2 \hat{q}^{0,1} = r(\lambda) \circ (\mathcal{D}_V^{0,1} + \lambda \mathcal{N}_V^{0,1} + (\lambda^2 - 1)q^{0,1}) \circ r(\lambda)^{-1},$$

when λ goes to 0, we conclude that

$$A_0^{0,1} = r(0) \circ (\mathcal{D}_V^{0,1} - q^{0,1}) \circ r(0)^{-1}$$

and, therefore, that

$$(d_V^{\hat{\lambda},\hat{q}})^{0,1} = r(0) \circ (\mathcal{D}_V^{0,1} - q^{0,1}) \circ r(0)^{-1} + \lambda A_1^{0,1} + \lambda^2 \hat{q}^{0,1}.$$

As for

$$(d_V^{\hat{\lambda},\hat{q}})^{1,0} = r(\lambda) \circ (\mathcal{D}_V^{1,0} + \lambda^{-1}\mathcal{N}^{1,0} + (\lambda^{-2} - 1)q^{1,0}) \circ r(\lambda)^{-1},$$

which has a pole at $\lambda = 0$, we have, for λ away from 0,

$$\sum_{i\geq 1} \lambda^{-i} A_{-i}^{1,0} + A_0^{1,0} + \sum_{i\geq 1} \lambda^i A_i^{1,0} = r(\lambda) \circ (\mathcal{D}_V^{1,0} + \lambda^{-1}\mathcal{N}^{1,0} + (\lambda^{-2} - 1)q^{1,0}) \circ r(\lambda)^{-1}, \tag{2.23}$$

with $A_{-i}^{1,0} \in \Omega^1(o(\underline{\mathbb{C}}^{n+2}))$, for all $i \geq 1$. Considering limits of (2.23) when λ goes to infinity, shows that $A_i^{1,0} = 0$, for all $i \geq 1$, and that

$$A_0^{1,0} = r(\infty) \circ (\mathcal{D}_V^{1,0} - q^{1,0}) \circ r(\infty)^{-1}.$$

Multiplying then both members of equation (2.23) by λ^2 and considering limits when λ goes to 0, we conclude that $A_{-2}^{1,0} = \hat{q}^{1,0}$ and that $A_{-i}^{1,0} = 0$, for all $i \geq 3$, and, ultimately, that

$$(d_V^{\hat{\lambda},\hat{q}})^{1,0} = r(\infty) \circ (\mathcal{D}_V^{1,0} - q^{1,0}) \circ r(\infty)^{-1} + \lambda^{-1} A_{-1}^{1,0} + \lambda^{-2} \hat{q}^{1,0}.$$

Thus

$$\begin{aligned}\hat{d}_V^{\lambda,\hat{q}} &= r(0)\circ(\mathcal{D}_V^{0,1}-q^{0,1}+q^{1,0})\circ r(0)^{-1}\\ &\quad + r(\infty)\circ(\mathcal{D}_V^{1,0}-q^{1,0}+q^{0,1})\circ r(\infty)^{-1}\\ &\quad + \lambda^{-1}A_{-1}^{1,0}+\lambda A_1^{0,1}+(\lambda^{-2}-1)\hat{q}^{1,0}+(\lambda^2-1)\hat{q}^{0,1},\end{aligned}$$

for $\lambda\in\mathbb{C}\backslash\{0\}$, and, in particular,

$$\begin{aligned}\hat{d} =& r(0)\circ(\mathcal{D}_V^{0,1}-q^{0,1}+q^{1,0})\circ r(0)^{-1}+r(\infty)\circ(\mathcal{D}_V^{1,0}-q^{1,0}+q^{0,1})\circ r(\infty)^{-1}\\ &+A_{-1}^{1,0}+A_1^{0,1}.\end{aligned}$$

The fact that $r(0)$ and $r(\infty)$ (and so $r(0)^{-1}$ and $r(\infty)^{-1}$), as well as q, preserve V and $V^\perp$, together with the $\mathcal{D}_V$-parallelness of V and of $V^\perp$, shows that $\hat{d}-(A_{-1}^{1,0}+A_1^{0,1})$ preserves $\Gamma(V)$ and $\Gamma(V^\perp)$. On the other hand, equations (2.21) and (2.22) combine to give

$$\hat{d}_V^{-\lambda,\hat{q}} = \rho\circ\hat{d}_V^{\lambda,\hat{q}}\circ\rho^{-1},$$

for all $\lambda\in\mathbb{C}\backslash\{0\}$ away from the poles of r and then, by continuity, on all of $\mathbb{C}\backslash\{0\}$. The particular case of $\lambda=1$ gives $\rho(A_{-1}^{1,0}+A_1^{0,1})_{|V}=-(A_{-1}^{1,0}+A_1^{0,1})_{|V}$ and $\rho(A_{-1}^{1,0}+A_1^{0,1})_{|V^\perp}=-(A_{-1}^{1,0}+A_1^{0,1})_{|V^\perp}$, showing that

$$A_{-1}^{1,0}+A_1^{0,1}\in\Omega^1(V\wedge V^\perp).$$

We conclude that

$$r(0)\circ(\mathcal{D}_V^{0,1}-q^{0,1}+q^{1,0})\circ r(0)^{-1}+r(\infty)\circ(\mathcal{D}_V^{1,0}-q^{1,0}+q^{0,1})\circ r(\infty)^{-1}=\mathcal{D}_V^{\hat{d}} \tag{2.24}$$

and

$$A_{-1}^{1,0}=(\mathcal{N}_V^{\hat{d}})^{1,0},\ \ A_1^{0,1}=(\mathcal{N}_V^{\hat{d}})^{0,1},$$

completing the proof. □

The flatness of $d_V^{\lambda,q}$ for all $\lambda\in\mathbb{C}\backslash\{0\}$ establishes that of $\hat{d}_V^{\lambda,\hat{q}}$, for all non-zero λ away from the poles of r and then, by continuity, for all $\lambda\in\mathbb{C}\backslash\{0\}$. By Proposition 2.18, we conclude that V is $(\hat{q},\hat{d})$-perturbed harmonic. Suppose $1\in\mathrm{dom}(r)$. By Lemma 2.15, it follows that:

Theorem 2.19. [9] *$r(1)^{-1}V$ is a $\mathrm{Ad}_{r(1)^{-1}}\,\hat{q}$-perturbed harmonic bundle.*

Note that this transformation preserves the harmonicity condition.

A transformation on the level of constrained Willmore surfaces follows, with some extra condition, as we shall see next. Set

$$\hat{\Lambda}^{1,0}:=r(\infty)\Lambda^{1,0},\quad \hat{\Lambda}^{0,1}:=r(0)\Lambda^{0,1}$$

and

$$\hat{\Lambda}=\hat{\Lambda}^{1,0}\cap\hat{\Lambda}^{0,1}.$$

Suppose, furthermore, that

$$\det r(0)_{|_V} = \det r(\infty)_{|_V}. \tag{2.25}$$

Then:

Theorem 2.20. [9]$(r(1)^{-1}\hat{\Lambda}^{1,0}, r(1)^{-1}\hat{\Lambda}^{0,1})$ *is a* $\mathrm{Ad}_{r(1)^{-1}}\hat{q}$*-constrained Willmore surface admitting* $r(1)^{-1}V$ *as a central sphere congruence.*

Proof. First of all, note that, by (2.18), $\hat{q}^{i,j} \in \Omega^{i,j}(\wedge^2\hat{\Lambda}^{j,i})$ and, therefore,

$$(\mathrm{Ad}_{r(1)^{-1}}\hat{q})^{i,j} \in \Omega^{i,j}(\wedge^2 r(1)^{-1}\hat{\Lambda}^{j,i}),$$

for $i \neq j \in \{0,1\}$. In the light of Theorem 2.19, we are left to verify that $(r(1)^{-1}\hat{\Lambda}^{1,0}, r(1)^{-1}\hat{\Lambda}^{0,1})$ is a surface admitting $r(1)^{-1}V$ as a central sphere congruence.

The fact that $\Lambda^{1,0}$ and $\Lambda^{0,1}$ are rank 2 isotropic subbundles of V ensures that so are $\hat{\Lambda}^{1,0}$ and $\hat{\Lambda}^{0,1}$, as $r(0)$ and $r(\infty)$ are orthogonal transformations and preserve V. To see that $\hat{\Lambda}$ is rank 1, we use some well-known facts about the Grassmannian $\mathcal{G}_W$ of isotropic 2-planes in a complex 4-dimensional space W: it has two components, each an orbit of the special orthogonal group $SO(W)$, intertwined by the action of elements of $O(W)\backslash SO(W)$, and for which any element intersects any element of the other component in a line while distinct elements of the same component have trivial intersection. Since $\operatorname{rank}\Lambda = 1$, $\Lambda^{1,0}_p$ and $\Lambda^{0,1}_p$ lie in different components of $\mathcal{G}_{V_p}$ and the hypothesis (2.25) ensures that the same is true of $\hat{\Lambda}^{1,0}_p$ and $\hat{\Lambda}^{0,1}_p$, for all p.

We are left to verify that

$$\hat{d}^{1,0}\Gamma(\hat{\Lambda}) \subset \Omega^1(\hat{\Lambda}^{1,0}), \quad \hat{d}^{0,1}\Gamma(\hat{\Lambda}) \subset \Omega^1(\hat{\Lambda}^{0,1}) \tag{2.26}$$

and that (recall equation (2.19))

$$(\mathcal{N}_V^{\hat{d}})^{1,0}\hat{\Lambda}^{0,1} = 0 = (\mathcal{N}_V^{\hat{d}})^{0,1}\hat{\Lambda}^{1,0}. \tag{2.27}$$

Equation (2.27) forces $\mathcal{N}_V^{\hat{d}}\hat{\Lambda} = 0$, in which situation, (2.26) reads

$$(\mathcal{D}_V^{\hat{d}})^{1,0}\Gamma(\hat{\Lambda}) \subset \Omega^1(\hat{\Lambda}^{1,0}), \quad (\mathcal{D}_V^{\hat{d}})^{0,1}\Gamma(\hat{\Lambda}) \subset \Omega^1(\hat{\Lambda}^{0,1}),$$

which, in its turn, follows from

$$(\mathcal{D}_V^{\hat{d}})^{1,0}\Gamma(\hat{\Lambda}^{1,0}) \subset \Omega^1(\hat{\Lambda}^{1,0}), \quad (\mathcal{D}_V^{\hat{d}})^{0,1}\Gamma(\hat{\Lambda}^{0,1}) \subset \Omega^1(\hat{\Lambda}^{0,1}). \tag{2.28}$$

It is (2.27) and (2.28) that we shall establish.

First of all, note that, according to (2.24),

$$(\mathcal{D}_V^{\hat{d}})^{1,0} = r(\infty) \circ (\mathcal{D}_V^{1,0} - q^{1,0}) \circ r(\infty)^{-1} + \hat{q}^{1,0}$$

and

$$(\mathcal{D}_V^{\hat{d}})^{0,1} = r(0) \circ (\mathcal{D}_V^{0,1} - q^{0,1}) \circ r(0)^{-1} + \hat{q}^{0,1}.$$

Now $q^{1,0}$ takes values in $\Lambda \wedge \Lambda^{0,1}$, so $q^{1,0}\Lambda^{1,0} \subset \Lambda \subset \Lambda^{1,0}$, by the isotropy of $\Lambda^{1,0}$. On the other hand, since rank $\hat{\Lambda} = 1$, we have $\wedge^2\hat{\Lambda}^{0,1} = \hat{\Lambda} \wedge \hat{\Lambda}^{0,1}$ and, therefore, $\hat{q}^{1,0}\hat{\Lambda}^{1,0} \subset \hat{\Lambda} \subset \hat{\Lambda}^{1,0}$. Together with Lemma 2.11, this establishes the $(1,0)$-part of (2.28). A similar argument establishes the $(0,1)$-part of it.

Finally, we establish (2.27). According to Proposition 2.18,

$$\begin{aligned}(\mathcal{N}_V^{\hat{d}})^{1,0} &= \lim_{\lambda\to 0} \lambda((\hat{d}_V^{\lambda,\hat{q}})^{1,0} - (\mathcal{D}_V^{\hat{d}})^{1,0} - (\lambda^{-2}-1)\hat{q}^{1,0}) \\ &= \lim_{\lambda\to 0} \lambda((\hat{d}_V^{\lambda,\hat{q}})^{1,0} - \lambda^{-2}\hat{q}^{1,0}) \\ &= \lim_{\lambda\to 0} (r(\lambda)\circ(\lambda\,(d_V^{\lambda,q})^{1,0})\circ r(\lambda)^{-1} - \lambda^{-1}\mathrm{Ad}_{r(0)}q^{1,0}) \\ &= \mathrm{Ad}_{r(0)}\mathcal{N}_V^{1,0} + \lim_{\lambda\to 0}\frac{1}{\lambda}(\mathrm{Ad}_{r(\lambda)} - \mathrm{Ad}_{r(0)})q^{1,0}.\end{aligned}$$

so that

$$(\mathcal{N}_V^{\hat{d}})^{1,0} = \mathrm{Ad}_{r(0)}\mathcal{N}_V^{1,0} + \frac{d}{d\lambda}_{|\lambda=0}\mathrm{Ad}_{r(\lambda)}q^{1,0};$$

and, similarly,

$$\begin{aligned}(\mathcal{N}_V^{\hat{d}})^{0,1} &= \lim_{\lambda\to\infty} \lambda^{-1}((\hat{d}_V^{\lambda,\hat{q}})^{0,1} - (\mathcal{D}_V^{\hat{d}})^{0,1} - (\lambda^2-1)\hat{q}^{0,1}) \\ &= \mathrm{Ad}_{r(\infty)}\mathcal{N}_V^{0,1} + \lim_{\lambda\to\infty}(r(\lambda)\circ\lambda q^{0,1}\circ r(\lambda)^{-1} - \lambda\mathrm{Ad}_{r(\infty)}q^{0,1})\end{aligned}$$

and, therefore,

$$(\mathcal{N}_V^{\hat{d}})^{0,1} = \mathrm{Ad}_{r(\infty)}\mathcal{N}_V^{0,1} + \frac{d}{d\lambda}_{|\lambda=0}\mathrm{Ad}_{r(\lambda^{-1})}q^{0,1}.$$

Furthermore, by (2.3.2),

$$(\mathcal{N}_V^{\hat{d}})^{1,0} = \mathrm{Ad}_{r(0)}(\mathcal{N}^{1,0} + [r(0)^{-1}\frac{d}{d\lambda}_{|\lambda=0}r(\lambda), q^{1,0}]).$$

The centrality of V with respect to $(\Lambda^{1,0}, \Lambda^{0,1})$ establishes, in particular, $\mathcal{N}_V^{1,0}\Lambda^{0,1} = 0$, whilst the isotropy of $\Lambda^{0,1}$ ensures, in particular, that $q^{1,0}\Lambda^{0,1} = 0$. Hence

$$\mathrm{Ad}_{r(0)}(\mathcal{N}^{1,0} + r(0)^{-1}\frac{d}{d\lambda}_{|\lambda=0}r(\lambda)\,q^{1,0})\hat{\Lambda}^{0,1} = 0.$$

On the other hand, differentiation of $r(\lambda)^{-1} = \rho\, r(-\lambda)^{-1}\rho$, derived from equation (2.22), gives

$$-r(\lambda)^{-1}\frac{d}{dk}_{|k=\lambda}r(k)\,r(\lambda)^{-1} = \rho\, r(-\lambda)^{-1}\frac{d}{dk}_{|k=-\lambda}r(k)\,r(-\lambda)^{-1}\rho,$$

or, equivalently,

$$\rho\, r(\lambda)^{-1}\frac{d}{dk}_{|k=\lambda}r(k)\rho = -r(-\lambda)^{-1}\frac{d}{dk}_{|k=-\lambda}r(k)\,r(-\lambda)^{-1}\rho\, r(\lambda)\rho,$$

and, therefore, yet again by equation (2.22),

$$\rho\, r(\lambda)^{-1}\frac{d}{dk}_{|_{k=\lambda}} r(k)\rho = -r(-\lambda)^{-1}\frac{d}{dk}_{|_{k=-\lambda}} r(k).$$

Evaluation at $\lambda = 0$ shows then that

$$\rho\, r(0)^{-1}\frac{d}{d\lambda}_{|_{\lambda=0}} r(\lambda)\rho = -r(0)^{-1}\frac{d}{d\lambda}_{|_{\lambda=0}} r(\lambda).$$

Equivalently,

$$r(0)^{-1}\frac{d}{d\lambda}_{|_{\lambda=0}} r(\lambda) \in \Gamma(V \wedge V^{\perp}).$$

Since $qV^{\perp} = 0$, we conclude that

$$Ad_{r(0)}(q^{1,0} r(0)^{-1}\frac{d}{d\lambda}_{|_{\lambda=0}} r(\lambda))\hat{\Lambda}^{0,1} = 0$$

and, ultimately, that $(\mathcal{N}_V^{\hat{d}})^{1,0}\hat{\Lambda}^{0,1} = 0$. A similar argument near $\lambda = \infty$ establishes $(\mathcal{N}_V^{\hat{d}})^{0,1}\hat{\Lambda}^{1,0} = 0$, completing the proof. □

2.3.3 Bäcklund transformation

We now construct $r = r(\lambda)$ satisfying the hypothesis of the previous section. As the philosophy underlying the work of C.-L. Terng and K. Uhlenbeck [22] suggests, we consider linear fractional transformations. As we shall see, a two-step process will produce a desired r.

Given $\alpha \in \mathbb{C}\backslash\{-1, 0, 1\}$ and L a null line subbundle of $\underline{\mathbb{C}}^{n+2}$ such that, locally, $\rho L \cap L^{\perp} = \{0\}$, set

$$p_{\alpha,L}(\lambda) := I\begin{cases} \frac{\alpha-\lambda}{\alpha+\lambda} & \text{on } L \\ 1 & \text{on } (L\oplus\rho L)^{\perp} \\ \frac{\alpha+\lambda}{\alpha-\lambda} & \text{on } \rho L \end{cases}$$

and

$$q_{\alpha,L}(\lambda) := I\begin{cases} \frac{\lambda-\alpha}{\lambda+\alpha} & \text{on } L \\ 1 & \text{on } (L\oplus\rho L)^{\perp} \\ \frac{\lambda+\alpha}{\lambda-\alpha} & \text{on } \rho L \end{cases},$$

for $\lambda \in \mathbb{C}\backslash\{\pm\alpha\}$, defining in this way maps

$$p_{\alpha,L}, q_{\alpha,L} : \mathbb{C}\backslash\{\pm\alpha\} \to \Gamma(O(\underline{\mathbb{C}}^{n+2}))$$

that, clearly, extend holomorphically to $\mathbb{P}^1\backslash\{\pm\alpha\}$, by setting

$$p_{\alpha,L}(\infty) := I\begin{cases} -1 & \text{on } L \\ 1 & \text{on } (L\oplus\rho L)^{\perp} \\ -1 & \text{on } \rho L \end{cases}$$

and

$$q_{\alpha,L}(\infty) := I.$$

Obviously, $p_{\alpha,L}(\infty)$ and $q_{\alpha,L}(\infty)$ do not depend on α. For further reference, note that, for all $\lambda \in \mathbb{C}\backslash\{\pm\alpha, 0\}$, we have

$$p_{\alpha,L}(\lambda) = q_{\alpha^{-1},L}(\lambda^{-1}), \tag{2.29}$$

whilst

$$p_{\alpha,L}(0) = q_{\alpha,L}(\infty), \quad p_{\alpha,L}(\infty) = q_{\alpha,L}(0).$$

The isometry $\rho = \rho^{-1}$ intertwines L and ρL and, therefore, preserves $(L \oplus \rho L)^{\perp}$, which makes clear that $\rho \circ p_{\alpha,L}(\lambda)$ and $p_{\alpha,L}(\lambda)^{-1} \circ \rho$ coincide in L, ρL and $(L \oplus \rho L)^{\perp}$ and, therefore,

$$\rho\, p_{\alpha,L}(\lambda)\rho = p_{\alpha,L}(\lambda)^{-1} = p_{\alpha,L}(-\lambda), \tag{2.30}$$

and, similarly,

$$\rho\, q_{\alpha,L}(\lambda)\rho = q_{\alpha,L}(\lambda)^{-1} = q_{\alpha,L}(-\lambda), \tag{2.31}$$

for all λ - both $p_{\alpha,L}$ and $q_{\alpha,L}$ are twisted in the sense of Section 2.3.2.

Since ρL is not orthogonal to L, $\rho L \neq L$, so L is not a subbundle of V and, therefore, $\operatorname{rank} V \cap (L \oplus \rho L) = 1$. We conclude that

$$p_{\alpha,L}(\infty)_{|_V} = q_{\alpha,L}(0)_{|_V} = I \begin{cases} -1 & \text{on } V \cap (L \oplus \rho L) \\ 1 & \text{on } V \cap (L \oplus \rho L)^{\perp} \end{cases}$$

has determinant $-1 \neq 1 = \det p_{\alpha,L}(0)_{|_V} = \det q_{\alpha,L}(\infty)_{|_V}$, so we cannot take $p_{\alpha,L} = r$ or $q_{\alpha,L} = r$ in the analysis of Section 2.3.2. However, we will be able to take $r = q_{\beta,\hat{L}} p_{\alpha,L}$, for suitable β and $\hat{L}$, as we shall see.

Now choose $\alpha \in \mathbb{C}\backslash\{-1, 0, 1\}$ and L^{α} a $d_V^{\alpha,q}$-parallel null line subbundle of $\underline{\mathbb{C}}^{n+2}$ such that, locally,

$$\rho L^{\alpha} \cap (L^{\alpha})^{\perp} = \{0\}. \tag{2.32}$$

Such a bundle L^{α} can be obtained by $d_V^{\alpha,q}$-parallel transport of $l_p^{\alpha} \in \mathbb{C}^{n+2}$, with l_p^{α} null and non-orthogonal to $\rho_p l_p^{\alpha}$, for some $p \in \Sigma$.

Lemma 2.21. [9] *There exists a holomorphic extension of*

$$\lambda \mapsto d^{\lambda,q}_{p_{\alpha,L^{\alpha}}} := p_{\alpha,L^{\alpha}}(\lambda) \circ d_V^{\lambda,q} \circ p_{\alpha,L^{\alpha}}(\lambda)^{-1}$$

to $\lambda \in \mathbb{C}\backslash\{0\}$ *through metric connections on* $\underline{\mathbb{C}}^{n+2}$.

Proof. We prove holomorphicity at $\lambda = \alpha$. For $\lambda \in \mathbb{C}\backslash\{0, \alpha\}$, write

$$d_V^{\lambda,q} = d_V^{\alpha,q} + (\lambda - \alpha)A(\lambda),$$

with $\lambda \mapsto A(\lambda) \in \Omega^1(o(\underline{\mathbb{C}}^{n+2}))$ holomorphic. Decompose $d_V^{\alpha,q} = D + \beta$ according to the decomposition

$$\underline{\mathbb{C}}^{n+2} = (L^{\alpha} \oplus \rho L^{\alpha}) \oplus (L^{\alpha} \oplus \rho L^{\alpha})^{\perp}.$$

The fact that $d_V^{\alpha,q}$ is a metric connection establishes

$$d_V^{\alpha,q}\,\Gamma(\rho L^\alpha) \subset \Omega^1(\rho L^\alpha)^\perp,$$

as well as

$$d_V^{\alpha,q}\,\Gamma((L^\alpha \oplus \rho L^\alpha)^\perp) \subset \Omega^1(L^\alpha)^\perp,$$

in view of the $d_V^{\alpha,q}$-parallelness of L^α. By (2.32), we conclude that the 1-form $\beta \in \Omega^1((L^\alpha \oplus \rho L^\alpha) \wedge (L^\alpha \oplus \rho L^\alpha)^\perp)$ takes values in $L^\alpha \wedge (L^\alpha \oplus \rho L^\alpha)^\perp$. Hence

$$p_{\alpha,L^\alpha}(\lambda) \circ \beta \circ p_{\alpha,L^\alpha}(\lambda)^{-1} = \frac{\alpha-\lambda}{\alpha+\lambda}\,\beta.$$

On the other hand,

$$p_{\alpha,L^\alpha}(\lambda) \circ D \circ p_{\alpha,L^\alpha}(\lambda)^{-1} = D,$$

as L^α, ρL^α and $(L^\alpha \oplus \rho L^\alpha)^\perp$ are all D-parallel. Thus

$$d^{\lambda,q}_{p_{\alpha,L}} = D + \frac{\alpha-\lambda}{\alpha+\lambda}\,\beta + (\lambda-\alpha)\,p_{\alpha,L^\alpha}(\lambda)\,A(\lambda)\,p_{\alpha,L^\alpha}(\lambda)^{-1}.$$

Lastly, note that, by the skew-symmetry of $A(\lambda)$, we have $A(\lambda)L^\alpha \subset (L^\alpha)^\perp$ and $A(\lambda)\rho L^\alpha \subset (\rho L^\alpha)^\perp$ and, therefore, by (2.32),

$$A(\lambda)L^\alpha \subset L^\alpha \oplus (L^\alpha \oplus \rho L^\alpha)^\perp$$

and

$$A(\lambda)\rho L^\alpha \subset \rho L^\alpha \oplus (L^\alpha \oplus \rho L^\alpha)^\perp.$$

We conclude that $(\lambda-\alpha)\,\mathrm{Ad}_{p_{\alpha,L^\alpha}(\lambda)}\,A(\lambda)$ has at most a simple pole at $\lambda = -\alpha$ and, therefore, that $d^{\lambda,q}_{p_{\alpha,L}}$ is holomorphic at $\lambda = \alpha$. Furthermore, the fact that D is a metric connection establishes that so is $d^{\lambda,q}_{p_{\alpha,L}}$, in view of the skew-symmetry of $A(\lambda)$ and of β.

Holomorphicity at $\lambda = -\alpha$ can either be proved in the same way, having in consideration that the $d_V^{\alpha,q}$-parallelness of L^α establishes the $d_V^{-\alpha,q}$-parallelness of ρL^α, or by exploiting the symmetry $\lambda \mapsto -\lambda$. □

Remark 3. 1) *The same argument establishes the existence of a holomorphic extension of*

$$\lambda \mapsto d^{\lambda,q}_{q_{\alpha,L^\alpha}} := q_{\alpha,L^\alpha}(\lambda) \circ d_V^{\lambda,q} \circ q_{\alpha,L^\alpha}(\lambda)^{-1}$$

to $\lambda \in \mathbb{C}\backslash\{0\}$ *through metric connections on* $\underline{\mathbb{C}}^{n+2}$.

2) *This argument uses nothing about the precise form of* $d_V^{\lambda,q}$, *only that it is holomorphic near* $\lambda = \pm\alpha$.

Now we can iterate the procedure starting with the connections $d^{\lambda,q}_{p_{\alpha,L^\alpha}}$. Choose $\beta \neq \pm\alpha$ in $\mathbb{C}\backslash\{-1,0,1\}$ and L^β a $d^{\beta,q}_V$-parallel null line subbundle of $\underline{\mathbb{C}}^{n+2}$. The fact that

$$p_{\alpha,L^\alpha} : (\underline{\mathbb{C}}^{n+2}, d^{\lambda,q}_V) \to (\underline{\mathbb{C}}^{n+2}, d^{\lambda,q}_{p_{\alpha,L^\alpha}})$$

preserves connections establishes the $d^{\beta,q}_{p_{\alpha,L^\alpha}}$-parallelness of

$$\hat{L}^\beta_\alpha := p_{\alpha,L^\alpha}(\beta)L^\beta.$$

Choose L^β satisfying, furthermore, $\rho\hat{L}^\beta_\alpha \cap (\hat{L}^\beta_\alpha)^\perp = \{0\}$. Such a bundle L^β can be obtained by $d^{\beta,q}_V$-parallel transport of $l^\alpha_p \in L^\alpha_p$, with l^α_p non-zero, non-orthogonal to $\rho_p l^\alpha_p$, for some $p \in \Sigma$. Indeed, by (2.30),

$$\begin{aligned}(\rho_p p_{\alpha,L^\alpha_p}(\beta)l^\alpha_p, p_{\alpha,L^\alpha_p}(\beta)l^\alpha_p) &= (p_{\alpha,L^\alpha_p}(\beta)^{-1}\rho_p l^\alpha_p, p_{\alpha,L^\alpha_p}(\beta)l^\alpha_p)\\ &= \frac{(\alpha-\beta)^2}{(\alpha+\beta)^2}(\rho_p l^\alpha_p, l^\alpha_p).\end{aligned}$$

It follows that

$$\lambda \mapsto q_{\beta,\hat{L}^\beta_\alpha}(\lambda)\, p_{\alpha,L^\alpha}(\lambda) \circ d^{\lambda,q}_V \circ p_{\alpha,L^\alpha}(\lambda)^{-1}\, q_{\beta,\hat{L}^\beta_\alpha}(\lambda)^{-1}$$

admits a holomorphic extension to $\lambda \in \mathbb{C}\backslash\{0\}$ through metric connections on $\underline{\mathbb{C}}^{n+2}$ and, furthermore, that

$$r^* := q_{\beta,\hat{L}^\beta_\alpha} p_{\alpha,L^\alpha}$$

satisfies all the hypothesis of Section 2.3.2 on r. Set

$$q^* := \mathrm{Ad}_{r^*(1)^{-1}}(\mathrm{Ad}_{r^*(0)}q^{1,0} + \mathrm{Ad}_{r^*(\infty)}q^{0,1}).$$

Definition 2.22. *The q^*-perturbed harmonic bundle*

$$V^* := r^*(1)^{-1}V$$

is said to be the Bäcklund transform of V of parameters $\alpha, \beta, L^\alpha, L^\beta$. *The q^*-constrained Willmore surface*

$$(\Lambda^{1,0}, \Lambda^{0,1})^* := (r^*(1)^{-1}r^*(\infty)\Lambda^{1,0}, r^*(1)^{-1}r^*(0)\Lambda^{0,1})$$

is said to be the Bäcklund transform of $(\Lambda^{1,0}, \Lambda^{0,1})$ of parameters $\alpha, \beta, L^\alpha, L^\beta$.

Note that transformations corresponding to the zero multiplier preserve the class of Willmore surfaces.

For further reference, set

$$((\Lambda^*)^{1,0}, (\Lambda^*)^{0,1}) := (r^*(1)^{-1}r^*(\infty)\Lambda^{1,0}, r^*(1)^{-1}r^*(0)\Lambda^{0,1}).$$

2.3.3.1 Bianchi permutability

Next we establish a *Bianchi permutability* of type p and type q transformations, showing that starting the procedure above with the connections $d^{\lambda,q}_{q_{\beta,L^\beta}}$ (when defined), instead of $d^{\lambda,q}_{p_{\alpha,L^\alpha}}$, produces the same transforms. The underlying argument will play a crucial role when investigating the preservation of reality conditions by Bäcklund transformation, in the next section.

Suppose $\rho L^\beta \cap (L^\beta)^\perp = \{0\}$ and set $\tilde{L}^\alpha_\beta := q_{\beta,L^\beta}(\alpha)L^\alpha$. Suppose, furthermore, that $\rho\tilde{L}^\alpha_\beta \cap (\tilde{L}^\alpha_\beta)^\perp = \{0\}$ (this is certainly the case for L^β obtained by $d^{\beta,q}_V$-parallel transport of $l^\alpha_p \in L^\alpha_p$, with l^α_p non-zero, non-orthogonal to $\rho_p l^\alpha_p$, for some $p \in \Sigma$). Analogously to r^*, we verify that

$$\hat{r}^* := p_{\alpha,\tilde{L}^\alpha_\beta} q_{\beta,L^\beta}$$

satisfies all the hypothesis of Section 2.3.2 on r. The next result, relating $\hat{r}^*$ to r^*, will be crucial in all that follows.

Lemma 2.23. [9]

$$\hat{r}^* = K\,r^*, \tag{2.33}$$

for $K := q_{\beta,L^\beta}(0)\, q_{\beta,\hat{L}^\beta_\alpha}(0)$.

The proof of the lemma we present next will be based on the following result:

Lemma 2.24. [4] *Let* $\gamma(\lambda) = \lambda\,\pi_{L_1} + \pi_{L_0} + \lambda^{-1}\,\pi_{L_{-1}}$ *and* $\hat{\gamma}(\lambda) = \lambda\,\pi_{\hat{L}_1} + \pi_{\hat{L}_0} + \lambda^{-1}\,\pi_{\hat{L}_{-1}}$ *be homomorphisms of* $\mathbb{C}^{n+2}$ *corresponding to decompositions*

$$\mathbb{C}^{n+2} = L_1 \oplus L_0 \oplus L_{-1} = \hat{L}_1 \oplus \hat{L}_0 \oplus \hat{L}_{-1}$$

with $L_{\pm 1}$ *and* $\hat{L}_{\pm 1}$ *null lines and* $L_0 = (L_1 \oplus L_{-1})^\perp$, $\hat{L}_0 = (\hat{L}_1 \oplus \hat{L}_{-1})^\perp$. *Suppose* $\mathrm{Ad}\,\gamma$ *and* $\mathrm{Ad}\,\hat{\gamma}$ *have simple poles. Suppose as well that* ξ *is a map into* $O(\mathbb{C}^{n+2})$ *holomorphic near* 0 *such that* $L_1 = \xi(0)\hat{L}_1$. *Then* $\gamma\xi\hat{\gamma}^{-1}$ *is holomorphic and invertible at* 0.

Now we proceed to the proof of Lemma 2.23.

Proof. For simplicity, write $p^{-1}_{\mu,L}$ and $q^{-1}_{\mu,L}$ for $\lambda \mapsto p_{\mu,L}(\lambda)^{-1}$ and, respectively, $\lambda \mapsto q_{\mu,L}(\lambda)^{-1}$, in the case $p_{\mu,L}$ and, respectively, $q_{\mu,L}$ are defined. As $L^\alpha = q_{\beta,L^\beta}(\alpha)^{-1}\tilde{L}^\alpha_\beta$, after an appropriate change of variable, we conclude, by Lemma 2.24, that $p_{\alpha,L^\alpha}\, q^{-1}_{\beta,L^\beta}\, p^{-1}_{\alpha,\tilde{L}^\alpha_\beta}$ admits a holomorphic and invertible extension to $\mathbb{P}^1\backslash\{\pm\beta, -\alpha\}$. On the other hand, in view of (2.30), the holomorphicity and invertibility of $p_{\alpha,L^\alpha}\, q^{-1}_{\beta,L^\beta}\, p^{-1}_{\alpha,\tilde{L}^\alpha_\beta}$ at the points α and $-\alpha$ are equivalent. Thus $p_{\alpha,L^\alpha}\, q^{-1}_{\beta,L^\beta}\, p^{-1}_{\alpha,\tilde{L}^\alpha_\beta}$ admits a holomorphic and invertible extension to $\mathbb{P}^1\backslash\{\pm\beta\}$,

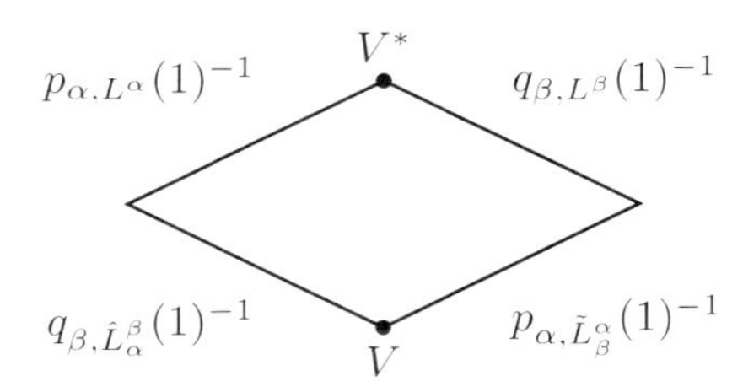

FIGURE 2.1: A Bianchi permutability of type p and type q transformations of perturbed harmonic bundles.

and so does, therefore, $(p_{\alpha,L^\alpha}\, q^{-1}_{\beta,L^\beta}\, p^{-1}_{\alpha,\tilde{L}^\alpha_\beta})^{-1}\, q^{-1}_{\beta,L^\beta}$. A similar argument shows that $p_{\alpha,\tilde{L}^\alpha_\beta}\,(q_{\beta,L^\beta}\, p^{-1}_{\alpha,L^\alpha}\, q^{-1}_{\beta,\hat{L}^\beta_\alpha})$ admits a holomorphic extension to $\mathbb{P}^1\backslash\{\pm\alpha\}$. But

$$p_{\alpha,\tilde{L}^\alpha_\beta}\, q_{\beta,L^\beta}\, p^{-1}_{\alpha,L^\alpha}\, q^{-1}_{\beta,\hat{L}^\beta_\alpha} = (p_{\alpha,L^\alpha}\, q^{-1}_{\beta,L^\beta}\, p^{-1}_{\alpha,\tilde{L}^\alpha_\beta})^{-1}\, q^{-1}_{\beta,L^\beta}.$$

We conclude that $p_{\alpha,\tilde{L}^\alpha_\beta}\, q_{\beta,L^\beta}\, p^{-1}_{\alpha,L^\alpha}\, q^{-1}_{\beta,\hat{L}^\beta_\alpha}$ extends holomorphically to $\mathbb{P}^1$ and is, therefore, constant. Evaluating at $\lambda = 0$ gives

$$p_{\alpha,\tilde{L}^\alpha_\beta}\, q_{\beta,L^\beta}\, p^{-1}_{\alpha,L^\alpha}\, q^{-1}_{\beta,\hat{L}^\beta_\alpha} = q_{\beta,L^\beta}(0)\, q_{\beta,\hat{L}^\beta_\alpha}(0),$$

completing the proof. □

According to (2.31), $\rho\, K\, \rho = K$, showing that K preserves V or, equivalently,

$$K\,V = V. \tag{2.34}$$

By (2.33), it follows that

$$r^{\,*}(1)^{-1}\, V = \hat{r}^{*}(1)^{-1}\, V,$$

establishing a *Bianchi permutability* of type p and type q transformations of perturbed harmonic bundles, by means of the commutativity of the diagram in Figure 2.1. Equation (2.33) makes clear, on the other hand, that

$$\hat{r}^{*}(1)^{-1}\, \hat{r}^{*}(\infty)\, \Lambda^{1,0} = r^{*}(1)^{-1}\, r^{*}(\infty)\, \Lambda^{1,0}$$

and

$$\hat{r}^{*}(1)^{-1}\, \hat{r}^{*}(0)\, \Lambda^{0,1} = r^{*}(1)^{-1}\, r^{*}(0)\, \Lambda^{0,1}.$$

We conclude that, despite not coinciding, r^* and $\hat{r}^*$ produce the same transforms of perturbed harmonic bundles and constrained Willmore surfaces. As a final remark, note that, yet again by equation (2.33),

$$\hat{q}^{*} := \mathrm{Ad}_{\hat{r}^*(1)^{-1}}(\mathrm{Ad}_{\hat{r}^*(0)} q^{1,0} + \mathrm{Ad}_{\hat{r}^*(\infty)} q^{0,1}) = q^{*}.$$

2.3.3.2 Real Bäcklund transformation

As we verify next, Bäcklund transformation preserves reality conditions, for special choices of parameters.

Suppose V is a real q-constrained harmonic bundle. Obviously, the reality of V establishes that of ρ and, therefore,

$$\overline{p_{\mu,L}(\lambda)} = p_{\overline{\mu},\overline{L}}(\overline{\lambda}), \qquad \overline{q_{\mu,L}(\lambda)} = q_{\overline{\mu},\overline{L}}(\overline{\lambda}), \tag{2.35}$$

for all μ, L and $\lambda \in \mathbb{C}\backslash\{\pm\mu\}$.

Lemma 2.25. *Suppose $\alpha \in \mathbb{C}\backslash(S^1 \cup \{0\})$. Then we can choose $\beta = \overline{\alpha}^{-1}$ and $L^\beta = \overline{L^\alpha}$ and both r^* and $\hat{r}^*$ are defined.*

Proof. The reality of ρ makes it clear that the non-orthogonality of L^α and ρL^α establishes that of $\overline{L^\alpha}$ and $\rho\overline{L^\alpha}$, as well as, together with (2.35) and (2.29), that, if

$$\rho p_{\alpha,L^\alpha}(\overline{\alpha}^{-1})\overline{L^\alpha} \cap p_{\alpha,L^\alpha}(\overline{\alpha}^{-1})\overline{L^\alpha} = \{0\},$$

then

$$\rho q_{\overline{\alpha}^{-1},\overline{L^\alpha}}(\alpha)L^\alpha \cap q_{\overline{\alpha}^{-1},\overline{L^\alpha}}(\alpha)L^\alpha = \{0\}.$$

On the other hand, the reality of V establishes that of $\mathcal{D}_V$ and $\mathcal{N}_V$, so that, by the reality of q,

$$d_V^{\overline{\alpha}^{-1},q} = \overline{d_V^{\alpha,q}}.$$

Hence the $d_V^{\alpha,q}$-parallelness of L^α establishes the $d_V^{\overline{\alpha}^{-1},q}$-parallelness of $\overline{L^\alpha}$. Obviously, if α is non-unit, then $\overline{\alpha}^{-1} \neq \pm\alpha$. We are left to verify that we can choose L^α a $d_V^{\alpha,q}$-parallel null line subbundle of $\underline{\mathbb{C}}^{n+2}$ such that, locally, $\rho L^\alpha \cap L^\alpha = \{0\}$ and

$$\rho p_{\alpha,L^\alpha}(\overline{\alpha}^{-1})\overline{L^\alpha} \cap p_{\alpha,L^\alpha}(\overline{\alpha}^{-1})\overline{L^\alpha} = \{0\}.$$

For this, let v and w be sections of V and $V^\perp$, respectively, with (v,v) never-zero, $(v,\overline{v}) = 0$ and $(w,w) = -(v,v)$. Define a null section of $\underline{\mathbb{C}}^{n+2}$ by $l^\alpha := v + w$ and then $L^\alpha \subset \underline{\mathbb{C}}^{n+2}$ by $d_V^{\alpha,q}$-parallel transport of l^α_p, for some point $p \in \Sigma$. $\square$

Let us focus then on the particular case of Bäcklund transformation of parameters $\alpha, \beta, L^\alpha, L^\beta$ with

$$\alpha \in \mathbb{C}\backslash(S^1 \cup \{0\}),\ \ \beta = \overline{\alpha}^{-1},\ \ L^\beta = \overline{L^\alpha},$$

which we refer to as *Bäcklund transformation of parameters α, L^α*. For this particular choice of parameters, we write $\tilde{L}^\alpha$ and $\hat{L}_\alpha$ for $\tilde{L}^\alpha_\beta$ and $\hat{L}^\beta_\alpha$, respectively. Note that, by (2.29) and (2.35), $\overline{\hat{L}_\alpha} = \tilde{L}^\alpha$. On the other hand,

$$\overline{r^*(1)^{-1}} = \overline{p_{\alpha,L^\alpha}(1)^{-1}}\,\overline{q_{\beta,\hat{L}_\alpha}(1)^{-1}} = q_{\beta,L^\beta}(1)^{-1}p_{\alpha,\overline{\hat{L}_\alpha}}(1)^{-1},$$

whilst, by (2.33),

$$r^*(1)^{-1} = (K^{-1}\hat{r}^*(1))^{-1} = q_{\beta,L^\beta}(1)^{-1}p_{\alpha,\tilde{L}^\alpha}(1)^{-1}K.$$

Hence

$$\overline{r^*(1)^{-1}} = r^*(1)^{-1}K^{-1}. \tag{2.36}$$

By (2.34), it follows that $\overline{V^*} = V^*$. Next we establish the reality of q^*. Yet again by (2.35),

$$\overline{r^*(0)} = q_{\overline{\alpha^{-1}},\overline{\hat{L}_\alpha}}(0) = p_{\overline{\alpha^{-1}},\overline{\hat{L}_\alpha}}(\infty)$$

and, on the other hand, by (2.33),

$$r^*(\infty) = K^{-1}p_{\alpha,\tilde{L}_\alpha}(\infty),$$

so that

$$\overline{r^*(0)} = Kr^*(\infty). \tag{2.37}$$

Together with (2.36), this makes clear that $\overline{(q^*)^{1,0}} = (q^*)^{0,1}$, the reality of q establishes that of q^*. We conclude that:

Theorem 2.26. [9] *If V is a real q-perturbed harmonic bundle, then the Bäcklund transform V^* of V, of parameters α, L^α, is a real q^*-perturbed harmonic bundle.*

A real transformation on the level of constrained Willmore surface follows. Equation (2.33) plays, yet again, a crucial role, by showing that

$$r^*(1) = K^{-1}q_{\alpha^{-1},\tilde{L}^\alpha}(1)p_{\overline{\alpha},L^\beta}(1) = K^{-1}\overline{q_{\beta,\hat{L}_\alpha}(1)p_{\alpha,L^\alpha}(1)} = K^{-1}\overline{r^*(1)},$$

and, therefore,

$$\overline{r^*(1)^{-1}} = r^*(1)^{-1}\overline{K}. \tag{2.38}$$

Suppose $(\Lambda^{1,0}, \Lambda^{0,1})$ is a real surface, so that, in particular, $\overline{\Lambda^{1,0}} = \Lambda^{0,1}$. By (2.37) and (2.38), it follows that $\overline{(\Lambda^*)^{1,0}} = (\Lambda^*)^{0,1}$, establishing the reality of the bundle

$$\Lambda^* := (\Lambda^*)^{1,0} \cap (\Lambda^*)^{0,1}.$$

We conclude that:

Theorem 2.27. [9]*If Λ is a real q-constrained Willmore surface, then the Bäcklund transform Λ^* of Λ, of parameters α, L^α, is a real q^*-constrained Willmore surface.*

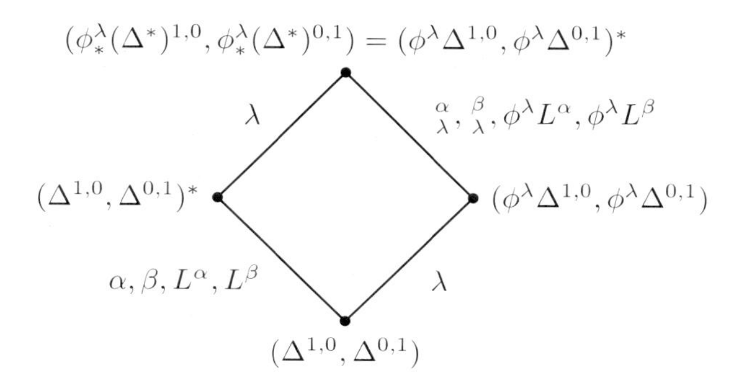

FIGURE 2.2: A Bianchi permutability of spectral deformation and Bäcklund transformation of constrained Willmore surfaces.

2.3.4 Spectral deformation versus Bäcklund transformation

Bäcklund transformation and spectral deformation permute, as follows:

Theorem 2.28. [9] *Let α,β,L^α,L^β be Bäcklund transformation parameters to V, $\lambda \in \mathbb{C}\backslash\{0, \pm\alpha, \pm\beta\}$ and*

$$\phi^\lambda : (\underline{\mathbb{C}}^{n+2}, d_V^{\lambda,q}) \to (\underline{\mathbb{C}}^{n+2}, d)$$

be an isometry of bundles preserving connections. The Bäcklund transform of parameters $\frac{\alpha}{\lambda}$, $\frac{\beta}{\lambda}$, $\phi^\lambda L^\alpha$, $\phi^\lambda L^\beta$ of the spectral deformation $\phi^\lambda V$ of V, of parameter λ, corresponding to the multiplier q, coincides with the spectral deformation of parameter λ, corresponding to the multiplier q^*, of the Bäcklund transform of parameters $\alpha, \beta, L^\alpha, L^\beta$ of V. Furthermore, if

$$\phi_*^\lambda : (\underline{\mathbb{C}}^{n+2}, d_{V^*}^{\lambda,q^*}) \to (\underline{\mathbb{C}}^{n+2}, d)$$

is an isometry preserving connections, then the diagram in Figure 2.2 commutes.

Proof. It is trivial, noting that $\phi^\lambda r^*(\lambda)^{-1} r^*(1) : (\underline{\mathbb{C}}^{n+2}, d_{V^*}^{\lambda,q^*}) \to (\underline{\mathbb{C}}^{n+2}, d)$ is an isometry of bundles preserving connections. □

For $\lambda \in \{\pm\alpha, \pm\beta\}$, it is not clear how the spectral deformation of parameter λ relates to the Bäcklund transformation of parameters $\alpha, \beta, L^\alpha, L^\beta$.

2.3.5 Isothermic surfaces under constrained Willmore transformation

The isothermic surface condition is known to be preserved under constrained Willmore spectral deformation:

Proposition 2.29. [8] *Constrained Willmore spectral deformation preserves the isothermic surface condition.*

Next we derive it in our setting.

Proof. Suppose (Λ, η) is an isothermic q-constrained Willmore surface, for some $\eta, q \in \Omega^1(\Lambda \wedge \Lambda^{(1)})$. Fix $\lambda \in S^1$ and $\phi_q^\lambda : (\underline{\mathbb{R}}^{n+1,1}, d_q^\lambda) \to (\underline{\mathbb{R}}^{n+1,1}, d)$ an isometry preserving connections. Set

$$\eta_\lambda := \lambda^{-1}\eta^{1,0} + \lambda\,\eta^{0,1}.$$

To prove the theorem, we show that $(\phi_q^\lambda\Lambda, \mathrm{Ad}_{\phi_q^\lambda}\eta_\lambda)$ is isothermic. Obviously, the reality of η establishes that of $\mathrm{Ad}_{\phi_q^\lambda}\eta_\lambda$. Recall (2.20) to conclude that

$$(\phi_q^\lambda\Lambda)^{(1)} = \phi_q^\lambda\Lambda^{(1)}$$

and, therefore, that $\mathrm{Ad}_{\phi_q^\lambda}\eta_\lambda$ takes values in $\phi_q^\lambda\Lambda \wedge (\phi_q^\lambda\Lambda)^{(1)}$. According to (2.12), we have

$$[q^{1,0} \wedge \eta^{0,1}] = 0 = [q^{0,1} \wedge \eta^{1,0}]$$

and, therefore,

$$\begin{aligned} d^{d_q^\lambda}\eta_\lambda &= d^{\mathcal{D}}\eta_\lambda + [(\lambda^{-1}\mathcal{N}^{1,0} + \lambda\mathcal{N}^{0,1} + (\lambda^{-2}-1)q^{1,0} + (\lambda^2-1)q^{0,1}) \wedge \eta_\lambda] \\ &= d^{\mathcal{D}}\eta_\lambda + [\mathcal{N} \wedge \eta]. \end{aligned}$$

According to the decomposition (2.13), we conclude that

$$d(\mathrm{Ad}_{\phi_q^\lambda}\eta_\lambda) = \phi_q^\lambda \circ d^{d_q^\lambda}\eta_\lambda \circ (\phi_q^\lambda)^{-1}$$

vanishes if and only if $d^{\mathcal{D}}\eta_\lambda = 0 = [\mathcal{N} \wedge \eta]$. Remark 1 and Lemma 2.5 complete the proof. □

As for Bäcklund transformation of isothermic constrained Willmore surfaces, we believe it does not necessarily preserve the isothermic condition. This shall be the subject of further work.

A very important subclass of isothermic constrained Willmore surfaces is the class of constant mean curvature surfaces in 3-dimensional space-forms. The constancy of the mean curvature of a surface in 3-dimensional space-form is preserved by both constrained Willmore spectral deformation, cf. [8], and constrained Willmore Bäcklund transformation, cf. [16], for special choices of parameters, with preservation of both the space-form and the mean curvature in the latter case. However, constant mean curvature surfaces are not conformally-invariant objects, requiring that we carry a distinguished space-form. This shall be the subject of a forthcoming paper. See [16, Section 8.2] and [17] for further details.

Acknowledgments

The author would like to thank Professor Magdalena Toda for the invitation to contribute an article to this monograph, which came as a great pleasure. Most of the underlying research work was carried out as part of the PhD studies of the author at the University of Bath, UK, under the supervision of Professor Francis Burstall, to whom the author is deeply grateful. The author was supported by Fundação para a Ciência e a Tecnologia, Portugal, with a PhD scholarship, and by Fundação da Faculdade de Ciências da Universidade de Lisboa, with a postdoctoral scholarship.

Bibliography

[1] W. Blaschke, *Vorlesungen über Differentialgeometrie III: Differentialgeometrie der Kreise und Kugeln*, Grundlehren XXIX, Springer, Berlin (1929).

[2] C. Bohle, G. P. Peters and U. Pinkall, *Constrained Willmore Surfaces*, Calculus of Variations and Partial Differential Equations 32 (2008), 263-277.

[3] R. Bryant, *A duality theorem for Willmore surfaces*, Journal of Differential Geometry 20 (1984), 23-53.

[4] F. E. Burstall, *Isothermic surfaces: Conformal geometry, Clifford algebras and Integrable systems*, Integrable systems, Geometry and Topology (ed. C.-L. Terng), Volume 36, American Mathematical Society/International Press Studies in Advanced Mathematics, American Mathematical Society, Providence (2006), 1-82.

[5] F. E. Burstall and D. M. J. Calderbank, *Conformal Submanifold Geometry I-III*, arXiv:1006.5700v1 [math.DG] (2010).

[6] F. E. Burstall, N. M. Donaldson, F. Pedit and U. Pinkall, *Isothermic Submanifolds of Symmetric R-Spaces*, Journal für die reine und angewandte Mathematik 2011, no. 660 (2011), 191-243.

[7] F. E. Burstall, D. Ferus, K. Leschke, F. Pedit and U. Pinkall, *Conformal Geometry of Surfaces in S^4 and Quaternions*, Lecture Notes in Mathematics, vol. 1772, Springer Verlag, Heidelberg, Berlin, New York, 2002.

[8] F. E. Burstall, F. Pedit and U. Pinkall, *Schwarzian Derivatives and Flows of Surfaces*, Contemporary Mathematics 308 (2002), 39-61.

[9] F. E. Burstall and A. C. Quintino, *Dressing transformations of constrained Willmore surfaces*, Communications in Analysis and Geometry 22 (2014), 469-518.

[10] F. E. Burstall and J. Rawnsley, *Twistor theory for Riemannian symmetric spaces*, Lecture Notes in Mathematics 1424, Springer-Verlag (1990).

[11] F. E. Burstall and S. D. Santos, *Special Isothermic Surfaces of Type d*, Journal of the London Mathematical Society 85(2) (2012), 571-591.

[12] J. Darboux, *Leçons sur la Théorie Générale des Surfaces et les Applications Géometriques du Calcul Infinitésimal, Parts 1 and 2*, Gauthier-Villars, Paris, 1887.

[13] N. Ejiri, *Willmore Surfaces with a Duality in $S^n(1)$*, Proceedings of the London Mathematical Society (3), 57(2) (1988), 383-416.

[14] S. Germain, *Recherches sur la théorie des surfaces élastiques*, Courcier, Paris (1821).

[15] S. Germain, *Remarques sur la nature, les bornes et l'etendue de la question des surfaces élastiques, et equation generale de ces surfaces*, Courcier, Paris (1826).

[16] A. C. Quintino, *Constrained Willmore Surfaces: Symmetries of a Möbius Invariant Integrable System* - Based on the author's PhD Thesis. arXiv:0912.5402v1 [math.DG] (2009).

[17] A. C. Quintino and S. D. Santos, *Polynomial conserved quantities for constrained Willmore surfaces* (in preparation). arXiv: 1507.01253v1 [mathDG] (2015).

[18] J. Richter, *Conformal Maps of a Riemann Surface into the Space of Quaternions*, PhD thesis, Technischen Universität Berlin (1997).

[19] M. Rigoli, *The Conformal Gauss Map of Submanifolds of the Möbius Space*, Annals of Global Analysis and Geometry 5(2) (1987), 97-116.

[20] C.-L. Terng and K. Uhlenbeck, *Bäcklund Transformations and Loop Group Actions*, Communications on Pure and Applied Mathematics 53 (2000), 1-75.

[21] G. Thomsen, Ueber konforme Geometrie I, *Grundlagen der Konformen Flaechentheorie*, Abhandlungen aus dem Mathematischen Seminar der Universität Hamburg 3 (1923), 31-56.

[22] K. Uhlenbeck, *Harmonic Maps into Lie Groups (Classical Solutons of the Chiral Model)*, Journal of Differential Geometry 30 (1989), 1-50.

[23] T. Willmore, *Note on Embedded Surfaces*, Analele Stiintifice ale Universitatii "Alexandru Ioan Cuza" din Iasi, N. Ser., Sect. Ia 11B (1965), 493-496.

[24] T. Willmore, *Riemannian Geometry*, Oxford Science Publications (1993).

Chapter 3

Analytical Representations of Willmore and Generalized Willmore Surfaces

Vassil M. Vassilev, Peter A. Djondjorov, Mariana Ts. Hadzhilazova, and Ivailo M. Mladenov

CONTENTS

Abstract

In this chapter we give analytical representations of Willmore and generalized Willmore (Helfrich) surfaces. All cylindrical surfaces of that type are described. Some special families of axially symmetric surfaces providing local extrema to the corresponding functionals are presented. The foregoing surfaces are determined analytically, the components of the respective position vectors being represented either in explicit form or by quadratures in terms of suitable variables. Parametric equations of Dupin cyclides and other Willmore surfaces obtained by inversions of catenoids and Enneper's surface are also presented.

3.1 Introduction

The functional

$$\mathcal{W} = \int_{\mathcal{S}} H^2 \mathrm{d}A \tag{3.1}$$

which assigns to each surface $\mathcal{S}$ immersed in the three-dimensional Euclidean space its total squared mean curvature H, $\mathrm{d}A$ being the induced surface element, was proposed about two centuries ago by the prominent French scientists Siméon Denis Poisson and Marie-Sophie Germain as the bending energy of thin elastic shells [10, 25]. It should be remarked also that an equivalent variational problem was studied about a century later, in 1923, by Thomsen [30] in connection with the conformal geometry. Nowadays, however, the functional (3.1) is widely known as the Willmore functional (energy) due to the work [38] published in 1965 by the English geometer Thomas James Willmore (see also [39]). The aforementioned note of T. J. Willmore attracted the interest of mathematicians to study the surfaces that provide local extrema to the functional (3.1) currently called "Willmore surfaces". These surfaces are determined by the solutions of the Euler-Lagrange equation associated with the Willmore functional, namely

$$\Delta_{\mathcal{S}} H + 2(H^2 - K)H = 0 \tag{3.2}$$

where $\Delta_{\mathcal{S}}$ is the Laplace-Beltrami operator on the surface $\mathcal{S}$ and K is its Gaussian curvature. It is noteworthy that the Willmore functional is invariant under the ten-parameter group of conformal transformations in $\mathbb{R}^3$ [36].
In 1973, Helfrich [14] had developed a mathematical model known nowadays as the spontaneous curvature model, to describe the equilibrium shapes of lipid bilayer membranes. Within the framework of this model, a generalization of the Willmore functional, namely

$$\mathcal{F}_b = \int_{\mathcal{S}} \left(\frac{1}{2} k_c (2H - \mathbb{h})^2 + k_G K \right) \mathrm{d}A \tag{3.3}$$

was introduced and interpreted as the bending (curvature) energy of the membrane; here k_c and k_G are the bending and Gaussian rigidities of the membrane, $\mathbb{h}$ is the so-called spontaneous curvature of the lipid bilayer.
Actually, in the Helfrich theory, the equilibrium shapes of a lipid bilayer membrane are determined by the critical points of the bending energy functional (3.3) subjected to the constraints of fixed total area A and enclosed volume V. Therefore, in this context, one is interested in the extremals of the functional

$$\mathcal{F} = \int_{\mathcal{S}} \left(\frac{1}{2} k_c (2H - \mathbb{h})^2 + k_G K \right) \mathrm{d}A + \lambda \int_{\mathcal{S}} \mathrm{d}A + p \int \mathrm{d}V. \tag{3.4}$$

Here λ and p are Lagrange multipliers corresponding to the constraints of fixed total area and enclosed volume, respectively. Let us remark that λ is interpreted as a tensile stress while p is seen as the hydrostatic pressure exerted on the membrane. The Euler-Lagrange equation corresponding to the functional (3.4) is derived by Ou-Yang and Helfrich and reads

$$\Delta_{\mathcal{S}} H + (2H + \text{Ih}) \left(H^2 - \frac{\text{Ih}}{2} H - K \right) - \frac{\lambda}{k_c} H = -\frac{p}{2k_c} \tag{3.5}$$

see [23].

In the past four decades, the Willmore surfaces were studied by many authors both from a purely mathematical point of view (see, e.g. [12, 26] and references therein) and in the context of applications, for instance, in structural mechanics, biophysics and mathematical biology (see, e.g. [3, 12, 16, 31, 32]). The same holds true for the Helfrich surfaces.

3.2 Cylindrical Surfaces

Let X, Y and Z be the axes of a right-handed rectangular Cartesian coordinate system $OXYZ$ centered at the origin O in $\mathbb{R}^3$. Consider a cylindrical surface $\mathcal{S}$ in $\mathbb{R}^3$, obtained by translating a plane curve Γ laying in the XOY-plane along the Z axis. In general, the components $\{x, y, z\}$ of the position vector of such a cylindrical surface read

$$\mathbf{x}(u, v) = \begin{pmatrix} x(u, v) \\ y(u, v) \\ z(u, v) \end{pmatrix} = \begin{pmatrix} x(v) \\ y(v) \\ u \end{pmatrix}, \qquad u, v \in \Omega \subseteq \mathbb{R}$$

where the functions $x(v)$ and $y(v)$ are assumed to have all continuous derivatives that we need in the domains where they are defined. The couple of functions $(x(v), y(v))$ represents the parametric equations of the directrix Γ of the surface $\mathcal{S}$. The first and second fundamental tensors of the surface $\mathcal{S}$ are

$$g_{\alpha\beta} = \begin{pmatrix} 1 & 0 \\ 0 & x_v^2 + y_v^2 \end{pmatrix}, \qquad b_{\alpha\beta} = \frac{1}{\sqrt{x_v^2 + y_v^2}} \begin{pmatrix} 0 & 0 \\ 0 & x_v y_{vv} - x_{vv} y_v \end{pmatrix}$$

respectively, and

$$g = \det(g_{\alpha\beta}) = x_v^2 + y_v^2, \quad \mathrm{d}A = \sqrt{g}\,\mathrm{d}u\mathrm{d}v = \sqrt{x_v^2 + y_v^2}\,\mathrm{d}u\mathrm{d}v. \tag{3.6}$$

Here and elsewhere, subscripts denote derivatives with respect to the corresponding variables. The mean and Gaussian curvatures of the regarded surfaces read

$$H = \frac{1}{2} \frac{x_v y_{vv} - x_{vv} y_v}{(x_v^2 + y_v^2)^{3/2}}, \qquad K = 0. \tag{3.7}$$

On account of equations (3.6) and (3.7), the bending energy (3.4) of a portion of a cylindrical surface $\mathcal{S}$ of unit height reads

$$\mathcal{F} = \int_{\Gamma} \mathcal{L} \mathrm{d}v \tag{3.8}$$

where

$$\mathcal{L} = \left(\frac{1}{2} k_c \left(\frac{x_v y_{vv} - x_{vv} y_v}{(x_v^2 + y_v^2)^{3/2}} - \mathbb{h} \right)^2 + \lambda - p \frac{x y_v - y x_v}{2\sqrt{x_v^2 + y_v^2}} \right) \sqrt{x_v^2 + y_v^2} \cdot$$

The application of the Euler operators

$$E_x = \frac{\partial}{\partial x} - D_v \frac{\partial}{\partial x_v} + D_v D_v \frac{\partial}{\partial x_{vv}} - \cdots$$

$$E_y = \frac{\partial}{\partial y} - D_v \frac{\partial}{\partial y_v} + D_v D_v \frac{\partial}{\partial y_{vv}} - \cdots$$

where

$$D_v = \frac{\partial}{\partial v} + x_v \frac{\partial}{\partial x} + y_v \frac{\partial}{\partial y} + x_{vv} \frac{\partial}{\partial x_v} + y_{vv} \frac{\partial}{\partial y_v} + \cdots$$

is the total differentiation operator with respect to the independent variable v on the Lagrangian density $\mathcal{L}$ of the functional (3.8) leads to the Euler-Lagrange equations $E_x \mathcal{L} = 0$ and $E_y \mathcal{L} = 0$. However, it turned out that $x_v E_x \mathcal{L} \equiv -y_v E_y \mathcal{L}$ and hence we get to an abnormal variational problem (see [22, Sec. 5.3]) having a single equation (3.5) determining the extremals of the Helfrich energy (3.8) instead of a system of two independent Euler-Lagrange equations for two dependent variables $x(v)$ and $y(v)$ as expected normally. In other words, we have to handle an underdetermined system and to complete it we may add another equation of our own choice to equation (3.5).

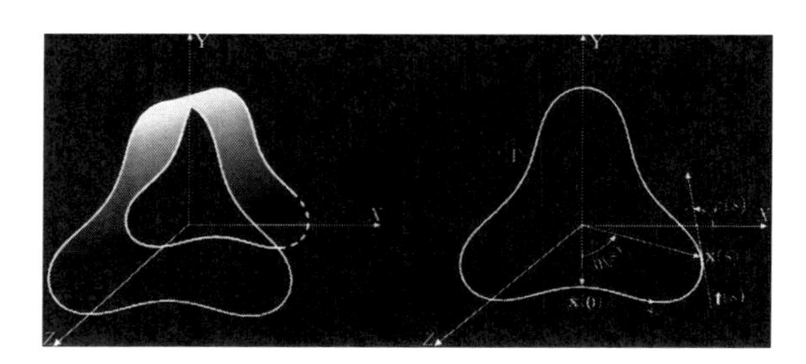

FIGURE 3.1: A slice of a cylindrical surface parametrized by arclength.

In this section we consider a complementary equation of the form

$$x_v^2 + y_v^2 = 1$$

which for any cylindrical surface $\mathcal{S}$ implies, first, that the independent variable v is, actually, the arclength s of its directrix Γ (see Figure 3.1) and hence $g_{\alpha\beta} = \delta_{\alpha\beta}$ where $\delta_{\alpha\beta}$ is the Kronecker delta symbol. Next, it shows that the coordinates $x(s)$ and $y(s)$ of such a directrix satisfy the equations

$$\dot{x}(s) = \cos\varphi(s), \qquad \dot{y}(s) = \sin\varphi(s) \tag{3.9}$$

where $\varphi(s)$ is the slope angle of the tangent vector to the directrix at the point $(x(s), y(s))$ (see Fig. 3.1) and hence, according to equation (3.7), the mean curvature of such a surface takes the form

$$H = \frac{1}{2}\kappa \tag{3.10}$$

where $\kappa = \dot{\varphi}$ is the curvature of the respective directrix Γ. Finally, substituting expressions (3.10) in equation (3.5) and taking into account that $g_{\alpha\beta} = \delta_{\alpha\beta}$ and $K = 0$ one gets to the following third-order nonlinear ordinary differential equation

$$2\dddot{\varphi} + \dot{\varphi}^3 - \mu\dot{\varphi} - \sigma = 0 \tag{3.11}$$

where $\mu = \mathrm{Ih}^2 + 2(\lambda/k_c)$, $\sigma = -2(p/k_c)$.
The three equations (3.9) and (3.11) describe entirely (up to a rigid motion in the XOY-plane) the directrices Γ of the cylindrical surfaces that determine critical points of the bending energy functional (3.8).
It is a simple matter to check that the following three conservation laws

$$2\ddot{\varphi}\cos\varphi + \dot{\varphi}^2\sin\varphi - \mu\sin\varphi - \sigma x = C_1 \tag{3.12}$$

$$2\ddot{\varphi}\sin\varphi - \dot{\varphi}^2\cos\varphi + \mu\cos\varphi - \sigma y = C_2 \tag{3.13}$$

$$\ddot{\varphi}^2 + \frac{1}{4}\dot{\varphi}^4 - \frac{1}{2}\mu\dot{\varphi}^2 - \sigma\dot{\varphi} = C_3 \tag{3.14}$$

hold on the smooth solutions of the system of equations (3.9) and (3.11). The first two of them allow explicit expressions of the parametric equations of the directrix Γ in terms of the curvature $\kappa = \dot{\varphi}$ and the slope angle φ. The third conservation law can be written in the form

$$\dot{\kappa}^2 = P[\kappa], \qquad P[\kappa] = -\frac{1}{4}\kappa^4 + \frac{1}{2}\mu\kappa^2 + \sigma\kappa + C_3. \tag{3.15}$$

It shows that equation (3.11) is integrable by quadratures and allows a representation of the curvature $\kappa(s) = \dot{\varphi}(s)$ and the slope angle $\varphi(s)$ of the regarded curves Γ in analytic form. Below, it is done in terms of the roots of the polynomial $P[\kappa]$.
Let us first stress that here we are interested in real-valued solutions $\kappa(s) \neq$ *const* of equation (3.15) possessing continuous derivatives at least up to the second order. Therefore, taking into account that μ, σ and C_3 are real numbers, it is clear that the corresponding polynomial $P[\kappa]$ should have at least two real roots, otherwise equation (3.15) does not have solutions of the considered type since the coefficient at the highest power κ^4 of $P[\kappa]$ is negative. Thus, only four alternative possibilities have to be considered, namely:

(a) the polynomial $P[\kappa]$ has two simple real roots $\alpha, \beta \in \mathbb{R}$, $\alpha < \beta$, and a pair of complex conjugate roots $\gamma, \delta \in \mathbb{C}$, $\delta = \bar{\gamma}$, which are both simple;

(b) the polynomial $P[\kappa]$ has four simple real roots $\alpha < \beta < \gamma < \delta \in \mathbb{R}$.

(c) the polynomial $P[\kappa]$ has two simple real roots $\alpha < \beta \in \mathbb{R}$ and one double real root $\gamma \in \mathbb{R}$.

(d) the polynomial $P[\kappa]$ has one simple real root $\alpha \in \mathbb{R}$ and one triple real root $\beta \in \mathbb{R}$.

Evidently, the polynomial $P[\kappa]$ is nonnegative in the intervals: $\alpha \leq \kappa \leq \beta$ in cases (a) and (c); $\alpha \leq \kappa \leq \beta$ and $\gamma \leq \kappa \leq \delta$ in case (b); $\alpha \leq \kappa \leq \beta$ or $\beta \leq \kappa \leq \alpha$ in case (d).

Using Vieta's formulas the coefficients μ, σ and C_3 of the polynomial $P[\kappa]$ can be expressed through its roots α, β, γ and δ as follows

$$\begin{aligned} \mu &= \frac{1}{2}\left(\alpha^2+\beta^2+\gamma^2+\alpha\beta+\alpha\gamma+\beta\gamma\right) \\ \sigma &= -\frac{1}{4}\left(\alpha+\beta\right)\left(\alpha+\gamma\right)\left(\beta+\gamma\right) \\ C_3 &= \frac{1}{4}\alpha\beta\gamma\left(\alpha+\beta+\gamma\right). \end{aligned} \tag{3.16}$$

Additionally, the absence of a term with κ^3 in the polynomial $P[\kappa]$ implies

$$\alpha+\beta+\gamma+\delta=0.$$

Now, we are in a position to express the curvature κ as a function of the arclength in terms of the roots of the polynomial $P[\kappa]$. Then, integrating, we may give an expression for the corresponding slope angle φ as well.

Theorem 3.1. *Given μ, σ and C_3, let the roots α, β, γ and δ of the respective polynomial $P[\kappa]$ be as in the case (a), that is $\alpha < \beta \in \mathbb{R}$, $\gamma, \delta \in \mathbb{C}$, $\delta = \bar{\gamma}$, and $\eta = (\gamma - \bar{\gamma})/2\mathrm{i} \neq 0$. Then the function*

$$\kappa_1(s) = \frac{(A\beta + B\alpha) - (A\beta - B\alpha)\,\mathrm{cn}(us,k)}{(A+B) - (A-B)\,\mathrm{cn}(us,k)} \tag{3.17}$$

where

$$A = \sqrt{4\eta^2 + (3\alpha+\beta)^2}, \qquad B = \sqrt{4\eta^2 + (\alpha+3\beta)^2}, \qquad u = \frac{1}{4}\sqrt{AB}$$

$$k = \frac{1}{\sqrt{2}}\sqrt{1 - \frac{4\eta^2 + (3\alpha+\beta)(\alpha+3\beta)}{\sqrt{\left[4\eta^2 + (3\alpha+\beta)(\alpha+3\beta)\right]^2 + 16\eta^2(\beta-\alpha)^2}}}$$

takes real values for each $s \in \mathbb{R}$ and satisfies equation (3.11). This function is periodic, its least period being $T_1 = (4/u)\,K(k)$, and its indefinite integral $\varphi_1(s)$ such that $\varphi_1(0) = 0$ is

$$\varphi_1(s) = \frac{A\beta - B\alpha}{A-B}s + \frac{(A+B)(\alpha-\beta)}{2u(A-B)}\Pi\left(-\frac{(A-B)^2}{4AB}, \mathrm{am}(us,k), k\right)$$

$$+\frac{\alpha-\beta}{2u\sqrt{k^2+\frac{(A-B)^2}{4AB}}}\arctan\left(\sqrt{k^2+\frac{(A-B)^2}{4AB}}\,\frac{\operatorname{sn}(us,k)}{\operatorname{dn}(us,k)}\right). \tag{3.18}$$

The *amplitude* function am$(\cdot)$ which appears as an argument in the above formula is defined by means of the equation (cf. Lawden [19])

$$\mathrm{am(u)}=\int_0^u \mathrm{dn}(v)\mathrm{d}v.$$

The elliptic integrals are defined in the Appendix.

Theorem 3.2. *Given μ, σ and C_3, let the roots α,β,γ and δ of the respective polynomial $P[\kappa]$ be as in the case (b), that is $\alpha<\beta<\gamma<\delta\in\mathbb{R}$. Consider the functions*

$$\kappa_2(s)=\delta-\frac{(\delta-\alpha)(\delta-\beta)}{(\delta-\beta)+(\beta-\alpha)\operatorname{sn}^2(us,k)} \tag{3.19}$$

$$\kappa_3(s)=\beta+\frac{(\gamma-\beta)(\delta-\beta)}{(\delta-\beta)-(\delta-\gamma)\operatorname{sn}^2(us,k)} \tag{3.20}$$

of the real variable s, where

$$u=\frac{1}{4}\sqrt{(\gamma-\alpha)(\delta-\beta)},\qquad k=\sqrt{\frac{(\beta-\alpha)(\delta-\gamma)}{(\gamma-\alpha)(\delta-\beta)}}.$$

Then, both functions (3.19) and (3.20) take real values for each $s\in\mathbb{R}$ and satisfy equation (3.11), they are periodic with least period $T_2=(2/u)K(k)$ and their indefinite integrals $\varphi_2(s)$ and $\varphi_3(s)$, respectively, such that $\varphi_2(0)=\varphi_3(0)=0$ are

$$\varphi_2(s) = \delta s-\frac{\delta-\alpha}{u}\Pi\left(\frac{\beta-\alpha}{\beta-\delta},\mathrm{am}(us,k),k\right) \tag{3.21}$$

$$\varphi_3(s) = \beta s-\frac{\beta-\gamma}{u}\Pi\left(\frac{\delta-\gamma}{\delta-\beta},\mathrm{am}(us,k),k\right). \tag{3.22}$$

Theorem 3.3. *Given μ, σ and C_3, let the roots of the respective polynomial $P[\kappa]$ be as in the case (c), i.e., $\alpha<\beta\in\mathbb{R}$, $\gamma=\delta\in\mathbb{R}$. Denote*

$$A=\sqrt{(3\alpha+\beta)^2},\qquad B=\sqrt{(\alpha+3\beta)^2}\qquad u=\frac{1}{4}\sqrt{AB}.$$

Then, if $(3\alpha+\beta)(\alpha+3\beta)>0$, the function

$$\kappa_4(s)=\frac{B\alpha+A\beta+(B\alpha-A\beta)\cos(us)}{B+A+(B-A)\cos(us)}$$

takes real values for each $s \in \mathbb{R}$ *and satisfies equation (3.11). This function is periodic with a least period* $T_4 = (4/u)\, K(k)$, *and its indefinite integral* $\varphi_4(s)$ *such that* $\varphi_4(0) = 0$ *is*

$$\varphi_4(s) = \frac{A\beta - B\alpha}{A - B} s + \frac{2\sqrt{A}\sqrt{B}\,(\alpha - \beta)}{u\,(A - B)} \arctan\left(\frac{\sqrt{A}}{\sqrt{B}} \tan\left(\frac{us}{2}\right)\right).$$

If $(3\alpha + \beta)(\alpha + 3\beta) < 0$, *then, the function*

$$\kappa_5(s) = \frac{B\alpha - A\beta + (B\alpha + A\beta)\cosh(us)}{B - A + (B + A)\cosh(us)}$$

takes real values for each $s \in \mathbb{R}$, *satisfies equation (3.11), and its indefinite integral* $\varphi_5(s)$ *such that* $\varphi_5(0) = 0$ *is*

$$\varphi_5(s) = \frac{A\beta + B\alpha}{A + B} s + \frac{2\sqrt{A}\sqrt{B}\,(\alpha - \beta)}{u\,(A + B)} \operatorname{arctanh}\left(\frac{\sqrt{\mathrm{A}}}{\sqrt{\mathrm{B}}} \tanh\left(\frac{\mathrm{us}}{2}\right)\right).$$

Theorem 3.4. *Given* μ, σ *and* C_3, *let the roots of the respective polynomial* $P[\kappa]$ *be as in the case (d), i.e., it has one simple real root* α *and one triple real root* β. *Then, the function*

$$\kappa_6(s) = \beta - \frac{4\beta}{1 + \beta^2 s^2}$$

satisfies equation (3.11) for each $s \in \mathbb{R}$ *and its indefinite integral* $\varphi_6(s)$ *such that* $\varphi_6(0) = 0$ *reads*

$$\varphi_6(s) = \beta s - 4 \arctan(\beta s).$$

As for the parametric equations of the regarded curves Γ, let us first consider the case $\sigma \neq 0$. Then, the conservation laws (3.12) and (3.13) can be cast in the form

$$\begin{aligned} x &= \tfrac{2}{\sigma}\dot{\kappa}\cos\varphi + \tfrac{1}{\sigma}\left(\kappa^2 - \mu\right)\sin\varphi - \tfrac{C_1}{\sigma} \\ y &= \tfrac{2}{\sigma}\dot{\kappa}\sin\varphi - \tfrac{1}{\sigma}\left(\kappa^2 - \mu\right)\cos\varphi - \tfrac{C_2}{\sigma} \end{aligned} \tag{3.23}$$

and represent the parametric equations of the directrix Γ in terms of the slope angle $\varphi(s)$ and its derivative, i.e., the curvature. Apparently, when $\sigma \neq 0$ the constants C_1 and C_2 correspond just to translations of the origin and therefore in this case, without loss of generality, we choose $C_1 = C_2 = 0$.
In the case $\sigma = 0$, the three conservation laws (3.12) – (3.14) take the form

$$2\ddot{\varphi}\cos\varphi + \dot{\varphi}^2\sin\varphi - \mu\sin\varphi = C_1 \tag{3.24}$$

$$2\ddot{\varphi}\sin\varphi - \dot{\varphi}^2\cos\varphi + \mu\cos\varphi = C_2 \tag{3.25}$$

$$\ddot{\varphi}^2 + \frac{1}{4}\dot{\varphi}^4 - \frac{1}{2}\mu\dot{\varphi}^2 = C_3. \tag{3.26}$$

It should be noted that the constant C_3 is determined through the roots of the polynomial $P(\dot{\varphi})$ by expression (3.16). Solving (3.24) – (3.25) for $\cos\varphi$ and $\sin\varphi$ one gets

$$\begin{aligned}\cos\varphi(s) &= \frac{2C_1\ddot{\varphi}(s) - C_2(\dot{\varphi}^2(s) - \mu)}{4C_3 + \mu^2} \\ \sin\varphi(s) &= \frac{2C_2\ddot{\varphi}(s) + C_1(\dot{\varphi}^2(s) - \mu)}{4C_3 + \mu^2}.\end{aligned} \tag{3.27}$$

Evidently, the sum of the squares of the right-hand sides of the above equations should be equal to one. Hence, on account of equation (3.26)

$$\frac{C_1^2 + C_2^2}{4C_3 + \mu^2} = 1.$$

Upon integration of equalities (3.27) one obtains

$$\begin{aligned}x(s) &= \frac{2C_1\dot{\varphi}(s) + C_2\mu s}{4C_3 + \mu^2} - \frac{C_2}{4C_3 + \mu^2}\int \dot{\varphi}^2(s)\mathrm{d}s \\ y(s) &= \frac{2C_2\dot{\varphi}(s) - C_1\mu s}{4C_3 + \mu^2} + \frac{C_1}{4C_3 + \mu^2}\int \dot{\varphi}^2(s)\mathrm{d}s.\end{aligned}$$

In the case of two real and two complex roots of the polynomial (3.15), $\sigma = 0$ implies $\alpha = -\beta < 0$, $A = B = \sqrt{\eta^2 + \alpha^2}$ and the expression for the curvature reads

$$\dot{\varphi}_1(s) = \alpha\,\mathrm{cn}(us, k), \qquad u = \frac{\sqrt{\eta^2 + \alpha^2}}{2}, \qquad k = \sqrt{\frac{\alpha^2}{\eta^2 + \alpha^2}}$$

meaning that the parametric equations of the directrix Γ read

$$\begin{aligned}x(s) &= \frac{2C_1\alpha\mathrm{cn}(us, k) - C_2\left(2uE(\mathrm{am}(us, k), k) - (\mu + \eta^2)s\right)}{4C_3 + \mu^2} \\ y(x) &= \frac{C_1\left(2uE(\mathrm{am}(us, k), k) - (\mu + \eta^2)s\right) + 2C_2\alpha\mathrm{cn}(us, k)}{4C_3 + \mu^2}.\end{aligned} \tag{3.28}$$

In the case of four real roots of the polynomial (3.15) and $\sigma = 0$ it is shown in [33] that the two expressions for the curvature $\dot{\varphi}$ differ only by a rigid motion in the plane $(s, \dot{\varphi})$. Thus, the only distinct solution of equation (3.11) in this case reads

$$\dot{\varphi}_s(s) = \alpha\,\mathrm{dn}(\tilde{u}s, \tilde{k}), \qquad \tilde{u} = \frac{\alpha^2}{2}, \qquad \tilde{k} = -\frac{\sqrt{\alpha^2 - \beta^2}}{\alpha}$$

and the parametric equations of the directrix Γ are

$$x(s) = \frac{2C_1\alpha\mathrm{dn}(\tilde{u}s, \tilde{k}) + C_2\left(2\alpha E(\mathrm{am}(\tilde{u}s, \tilde{k}), \tilde{k}) + \mu s\right)}{4C_3 + \mu^2}$$

$$y(s) = \frac{-C_1\left(2\alpha E(\mathrm{am}(\tilde{u}s,\tilde{k}),\tilde{k})+\mu s\right)+2C_2\alpha\mathrm{dn}(\tilde{u}s,\tilde{k})}{4C_3+\mu^2}.$$

The case $\mu = \sigma = 0$ corresponds to the prominent Euler elasticae. In this case

$$\gamma = \pm \mathrm{i}\alpha, \qquad \eta = \pm\alpha, \qquad u = \sqrt{\frac{\alpha^2}{2}}, \qquad k = \frac{1}{\sqrt{2}}$$

and the parametric equations of the directrix Γ simplify to

$$\begin{aligned} x(s) &= \frac{C_1\alpha\mathrm{cn}(us,k) - C_2\left(2uE(\mathrm{am}(us,k),k)-\alpha^2 s\right)}{2C_3} \\ y(x) &= \frac{C_1\left(2uE(\mathrm{am}(us,k),k)-\alpha^2 s\right)+C_2\alpha\mathrm{cn}(us,k)}{2C_3}. \end{aligned} \tag{3.29}$$

In this way, having obtained in explicit form the solutions of equation (3.11), i.e., the curvatures and the corresponding slope angles, we have completely determined in analytic form the corresponding directrices Γ (up to a rigid motion in the plane $\mathbb{R}^2$) through the parametric equations (3.23) for the case $\sigma \neq 0$, (3.28) and (3.29) for the case $\sigma = 0$.

Now we are interested in directrices Γ that close up smoothly meaning that there exists a value L of the arclength s such that $\mathbf{x}(0) = \mathbf{x}(L)$ and $\mathbf{t}(0) = \mathbf{t}(L)$ where $\mathbf{t}(s)$ is the tangent vector. The later property of such a smooth closed directrix Γ and the definition of the tangent vector imply that there exists an integer m such that $\varphi(L) = \varphi(0) + 2m\pi$ where $\varphi(s)$ is the corresponding slope angle. Since in the current section we choose $\varphi(0) = 0$, this equality simplifies to $\varphi(L) = 2m\pi$.

Under the above assumptions, expressions (3.23) show that

$$\begin{aligned} \mathbf{x}\,(0) &= \left(\frac{2}{\sigma}\ddot{\varphi}(0), -\frac{1}{\sigma}(\dot{\varphi}^{\,2}(0)-\mu)\right) \\ \mathbf{x}\,(L) &= \left(\frac{2}{\sigma}\ddot{\varphi}(L), -\frac{1}{\sigma}(\dot{\varphi}^{\,2}(L)-\mu)\right). \end{aligned}$$

Consequently, the equality $\mathbf{x}(0) = \mathbf{x}(L)$ yields

$$\ddot{\varphi}(L) = \ddot{\varphi}(0), \qquad \dot{\varphi}(L) = \pm\dot{\varphi}(0)$$

which, on account of relation (3.15), implies that L is a period of the function $\dot{\varphi}(s)$, that is $L = nT$ where n is a positive integer and T is the least period of the function $\dot{\varphi}(s)$. Since $\varphi(nT) = n\varphi(T)$, then $2m\pi = \varphi(L) = \varphi(nT) = n\varphi(T)$ and hence

$$\varphi\,(T) = \frac{2m\pi}{n}\cdot \tag{3.30}$$

Thus, in the cases when $\sigma \neq 0$, relation (3.30) is found to be a necessary condition for a directrix Γ to close up smoothly. Apparently, it is a sufficient condition as well.

Straightforward computations lead to the following explicit expressions

$$\varphi_1(T_1) = \frac{4(A\beta - B\alpha)}{u(A-B)}K(k) + 2\frac{(A+B)(\alpha-\beta)}{u(A-B)}\Pi\left(-\frac{(A-B)^2}{4AB}, k\right)$$

$$\varphi_2(T_2) = \frac{2\delta}{u}K(k) + 2\frac{\alpha-\delta}{u}\Pi\left(\frac{\alpha-\beta}{\delta-\beta}, k\right)$$

$$\varphi_3(T_2) = \frac{2\beta}{u}K(k) + 2\frac{\gamma-\beta}{u}\Pi\left(\frac{\gamma-\delta}{\beta-\delta}, k\right)$$

for the angles of form (3.18), (3.21) and (3.22), respectively. These expressions and the closure condition (3.30) allow us to determine whether a curve of curvature (3.17), (3.19) or (3.20) closes up smoothly or not.

An interesting property of the curves of curvatures $\dot{\varphi}_2(s)$ and $\dot{\varphi}_3(s)$ is observed – if one of these curves closes up smoothly, then so does the other one. To see this let us first note that the solutions $\dot{\varphi}_2(s)$ and $\dot{\varphi}_3(s)$ correspond to case (b) where the polynomial $P[\kappa]$ has four real roots. Without loss of generality, these roots can be written in the form

$$\alpha = -3q - v - 2w, \qquad \beta = q - v - 2w, \qquad \gamma = q - v + 2w, \qquad \delta = q + 3v + 2w$$

where q, v and w are three arbitrary positive real numbers. The main advantage of this parametrization is that it preserves the order of the roots of the polynomial $P[\kappa]$, i.e., $\alpha < \beta < \gamma < \delta$ for any choice of the parameters q, v and w. Using these parameters, it is easy to find that

$$\frac{\partial\psi}{\partial\alpha} = \frac{\partial\psi}{\partial\beta} = \frac{\partial\psi}{\partial\gamma} = 0, \qquad \psi = \varphi_3(T_2) - \varphi_2(T_2)$$

meaning that $\psi = \text{const}$. This constant can be determined by evaluating the function ψ for any values of the roots α, β and γ, say $\alpha = -3$, $\beta = -2$, $\gamma = -1$, that gives

$$\varphi_3(T_2) - \varphi_2(T_2) = 4\pi.$$

This relation and the closure condition (3.30) do imply that the foregoing two curves close up simultaneously.

In what concerns the applications (e.g., determination of vesicle shapes), of special interest are solutions to equation (3.11) that give rise to closed non-self-intersecting (simple) curves. A sufficient condition for this is $\mu < 0$ which is discussed in [2]. It is also mentioned therein that the closed curves satisfying condition (3.30) with $m \neq \pm 1$ or $n = 1$ are necessarily self-intersecting. In this section, the case $\mu > 0$ is considered and several new sufficient conditions are presented for a closed curve meeting a closure condition of form (3.30) with $m = \pm 1$ and $n \geq 2$ to be simple or not.

It is convenient to treat the problem of the self-intersection in terms of the magnitude $r(s)$ of the position vector $\mathbf{x}(s)$ and the angle $\theta(s)$ between the position vectors $\mathbf{x}(0)$ and $\mathbf{x}(s)$. The angle $\theta(s)$ is assumed to be positive if the vector $\mathbf{x}(0)$ should be rotated counterclockwise to become unidirectional with the vector $\mathbf{x}(s)$ and negative otherwise. Hence

$$\begin{aligned} x(s) &= -\mathrm{sgn}(y(0))r(s)\sin\theta(s) \\ y(s) &= \mathrm{sgn}(y(0))r(s)\cos\theta(s) \end{aligned}$$

and then, according to expressions (3.23)

$$\dot{\theta}(s) = \frac{\dot{\varphi}^2(s) - \mu}{\sigma r^2(s)}.$$

The next four statements (whose proofs can be found in [33]) clarify the matter whether a curve of curvature satisfying (3.11) is self-intersecting or not.

Theorem 3.5. *Let Γ be a smooth closed curve whose curvature $\dot{\varphi}(s)$ is a solution of equation (3.11) of form (3.17), (3.19) or (3.20). Let T be the least period of the function $\dot{\varphi}(s)$ and let the corresponding slope angle meets a closure condition of form (3.30) with $m = \pm 1$ and $n \geq 2$. Then, the curve Γ is self-intersecting if and only if there exists $s_0 \in (0, T/2)$ such that $\theta(s_0) = \theta(0)$ or $\theta(s_0) = \theta(T/2)$.*

Theorem 3.6. *Let Γ be a smooth closed curve whose curvature $\dot{\varphi}(s)$ is a solution of equation (3.11) of form (3.17), (3.19) or (3.20). Let T be the least period of the function $\dot{\varphi}(s)$ and let the corresponding slope angle meets a closure condition of form (3.30) with $m = \pm 1$ and $n \geq 2$. Then:*

i) *the curve Γ is simple if $\dot{\varphi}^2(s) - \mu \neq 0$ for $s \in [0, T/2]$;*

ii) *the curve Γ is self-intersecting if the equation $\dot{\varphi}^2(s) - \mu = 0$ has exactly one solution for $s \in [0, T/2]$.*

Corollary 1. *Under the assumptions of Theorem 3.6, the curve Γ is simple if its curvature $\dot{\varphi}(s)$ is of constant sign in the interval $s \in [0, T/2]$.*

Corollary 2. *Under the assumptions of Theorem 3.6, the curve Γ is self-intersecting if its curvature $\dot{\varphi}(s)$ is such that the respective polynomial $P(\dot{\varphi})$ has real roots.*

A typical example of a closed directrix and the corresponding cylindrical surface is shown in Fig. 3.2. The directrices of cylindrical surfaces of 2, 3 and 4 axes of symmetry are shown in Fig. 3.3 for solutions according to Theorem 3.1 and in Fig. 3.4 for solutions according to Theorem 3.2. The evolution of a directrix with σ up to the value at which some opposite points come into contact is presented in Fig. 3.5.

Let us remark that the parametric equations (3.23) describe the behaviour of a carbon nanotube subject to hydrostatic pressure [34] and are used to study the collapse of such tubes [9].

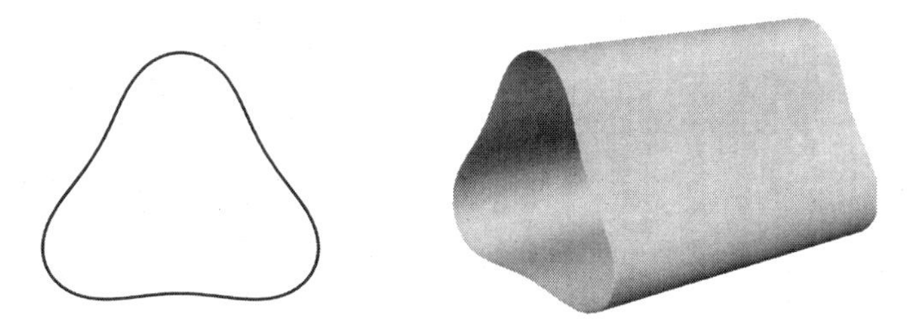

FIGURE 3.2: The directrix and the corresponding cylindrical surface.

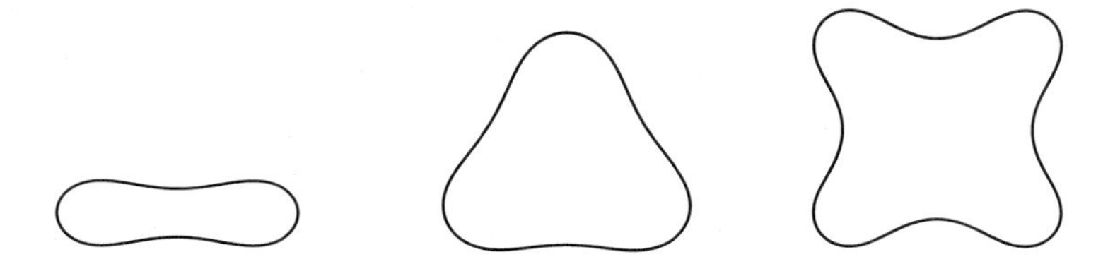

FIGURE 3.3: Directrices that are simple curves.

3.3 Axially Symmetric Surfaces

Let again X, Y, Z denote the axes of a right-handed rectangular Cartesian coordinate system $\{x, y, z\}$ in the three-dimensional Euclidean space $\mathbb{R}^3$. Without loss of generality, the components x, y, z of the position vector $\mathbf{x}$ of an axially symmetric surface $\mathcal{S}$ immersed in $\mathbb{R}^3$ can be given in the form

$$\mathbf{x}(u,v) = \begin{pmatrix} x(u,v) \\ y(u,v) \\ z(u,v) \end{pmatrix} = \begin{pmatrix} r(u)\cos v \\ r(u)\sin v \\ h(u) \end{pmatrix}, \quad u \in \Omega \subseteq \mathbb{R}, \quad v \in [0, 2\pi]$$

where the functions $r(u)$ and $h(u)$ are supposed to have as many derivatives as may be required on the domain Ω. Such a surface can be thought of as obtained by revolving around the OZ-axis a plane curve $\Gamma : u \mapsto (r(u), h(u))$ laying in the XOZ-plane.
In these notations, the first and second fundamental tensors of the surface $\mathcal{S}$ read

$$g_{\alpha\beta} = \begin{pmatrix} r_u^2 + h_u^2 & 0 \\ 0 & r^2 \end{pmatrix}, \quad b_{\alpha\beta} = \frac{1}{\sqrt{r_u^2 + h_u^2}} \begin{pmatrix} r_u h_{uu} - h_u r_{uu} & 0 \\ 0 & r h_u \end{pmatrix}$$

respectively, and

$$g = \det(g_{\alpha\beta}) = r^2\left(r_u^2 + h_u^2\right), \qquad \mathrm{d}A = \sqrt{g}\,\mathrm{d}u\mathrm{d}v = r\sqrt{r_u^2 + h_u^2}\,\mathrm{d}u\mathrm{d}v. \quad (3.31)$$

FIGURE 3.4: Self-intersecting directrices.

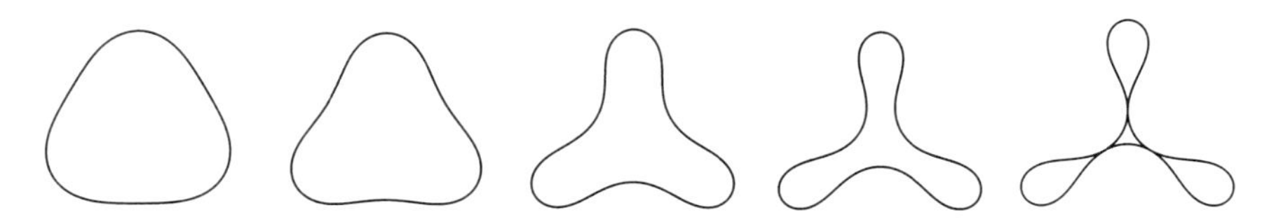

FIGURE 3.5: Evolution of the shape with the pressure.

Consequently, the mean and Gaussian curvatures of the regarded surface read

$$H = \frac{1}{2}\frac{r\left(r_u h_{uu} - h_u r_{uu}\right) + h_u\left(r_u^2 + h_{uu}^2\right)}{r\left(r_u^2 + h_u^2\right)^{3/2}}, \qquad K = \frac{h_u\left(r_u h_{uu} - h_u r_{uu}\right)}{r\left(r_u^2 + h_u^2\right)^2}. \tag{3.32}$$

On account of equations (3.31) and (3.32), the Willmore energy (3.1) of an axially symmetric surface $\mathcal{S}$ reads

$$\mathcal{W} = 2\pi \int_\Omega \mathcal{L} \mathrm{d}u \tag{3.33}$$

where

$$\mathcal{L} = r\sqrt{r_u^2 + h_u^2}\left(\frac{1}{2}\frac{r\left(r_u h_{uu} - h_u r_{uu}\right) + h_u\left(r_u^2 + h_{uu}^2\right)}{r\left(r_u^2 + h_u^2\right)^{3/2}}\right)^2.$$

The application of the Euler operators

$$E_r = \frac{\partial}{\partial r} - D_u\frac{\partial}{\partial r_u} + D_u D_u \frac{\partial}{\partial r_{uu}} - \cdots$$

$$E_h = \frac{\partial}{\partial h} - D_u\frac{\partial}{\partial h_u} + D_u D_u \frac{\partial}{\partial h_{uu}} - \cdots$$

where

$$D_u = \frac{\partial}{\partial u} + r_u\frac{\partial}{\partial r} + h_u\frac{\partial}{\partial h} + r_{uu}\frac{\partial}{\partial r_u} + h_{uu}\frac{\partial}{\partial h_u} + \cdots$$

is the total differentiation operator, on the Lagrangian density $\mathcal{L}$ of the functional (3.33) leads to the Euler-Lagrange equations $E_r\mathcal{L} = 0$ and $E_h\mathcal{L} = 0$. However, it turned out that $r_u E_r \mathcal{L} \equiv -h_u E_h \mathcal{L}$ and hence again we face an abnormal variational problem having a single equation (3.2) determining the extremals of the Willmore energy (3.33) instead of a system of two Euler-Lagrange equations for two dependent variables $r(u)$ and $h(u)$ as expected normally. In other words, we have got an under-determined system and to complete it we may add another equation of our own choice to equation (3.2). Four such cases are listed below.

Case (1). Assume $r = u$ and let $h_u = h_r = \tan\psi(r)$. In this case, equation (3.2) takes the form

$$\begin{aligned} \cos^3\psi\psi_{rrr} &= 4\sin\psi\cos^2\psi\psi_{rr}\psi_r - \tfrac{2\cos^3\psi}{r}\psi_{rr} \\ &\quad - \cos\psi\left(\sin^2\psi - \tfrac{1}{2}\cos^2\psi\right)\psi_r^3 + \tfrac{7\sin\psi\cos^2\psi}{2r}\psi_r^2 \\ &\quad - \left(\tfrac{\sin^2\psi - 2\cos^2\psi}{2r^2}\right)\cos\psi\psi_r - \left(\tfrac{\sin^2\psi + 2\cos^2\psi}{2r^2}\right)\tfrac{\sin\psi}{r} \end{aligned} \tag{3.34}$$

while the expressions (3.32) for the mean and Gaussian curvatures become

$$H = \frac{1}{2}\left(\cos\psi\psi_r + \frac{\sin\psi}{r}\right), \qquad K = \cos\psi\psi_r\frac{\sin\psi}{r}.$$

Obviously, here $\psi(r)$ represents the tangent angle. The following conservation law

$$\psi_{rr} - \frac{1}{2}\tan\psi\psi_r^2 + \frac{1}{r}\psi_r - \frac{\left(1+\cos^2\psi\right)\tan\psi}{2r^2\cos^2\psi} = \frac{c_1}{r\cos^3\psi}$$

holds on the solutions of equation (3.34) where c_1 is a constant of integration.

Case (2). Assume

$$r_u^2 + h_u^2 = 1.$$

This means that u is the arc length s along the contour curve Γ. Then, in terms of the tangent angle φ we have

$$\dot{r} = \cos\varphi, \qquad \dot{h} = \sin\varphi$$

and we can rewrite equation (3.2) in the form

$$\dddot{\varphi} = -\frac{2\cos\varphi}{r}\ddot{\varphi} - \frac{1}{2}\dot{\varphi}^3 + \frac{3\sin\varphi}{2r}\dot{\varphi}^2 + \frac{2-3\sin^2\varphi}{2r^2}\dot{\varphi} - \frac{\left(\cos^2\varphi+1\right)\sin\varphi}{2r^3} \tag{3.35}$$

where the dots indicate derivatives with respect to the arc length s. At the same time, the expressions (3.32) for the mean and Gaussian curvatures take the familiar form

$$H = \frac{1}{2}\left(\dot{\varphi} + \frac{\sin\varphi}{r}\right), \qquad K = \dot{\varphi}\frac{\sin\varphi}{r}.$$

The following conservation law

$$\ddot{\varphi} + \frac{1}{2}\tan\varphi\dot{\varphi}^2 + \frac{\cos\varphi}{r}\dot{\varphi} - \frac{\left(1+\cos^2\varphi\right)\tan\varphi}{2r^2} = \frac{c_2}{r\cos\varphi}$$

holds on the solutions of equation (3.35) where c_2 is a constant of integration.

Case (3). Assume $h = u$ and hence $r = r(h)$. Then, equation (3.2) becomes

$$\begin{aligned}
&2r^3\left(r_h^2+1\right)^2 r_{hhhh}\\
&+4r^2 r_h\left(r_h^2+1\right)\left(r_h^2+1-5rr_{hh}\right)r_{hhh}\\
&+5r^3\left(6r_h^2-1\right)r_{hh}^3-3r^2\left(r_h^2+1\right)\left(4r_h^2-1\right)r_{hh}^2\\
&-r\left(r_h^2+1\right)^2\left(2r_h^2-1\right)r_{hh}-\left(r_h^2+1\right)^3\left(2r_h^2+1\right)=0. \qquad (3.36)
\end{aligned}$$

and the mean and Gaussian curvatures read

$$H=\frac{1+r_h^2-rr_{hh}}{2r\left(1+r_h^2\right)^{3/2}}, \qquad K=-\frac{r_{hh}}{r\left(1+r_h^2\right)^2}.$$

The conservation law

$$r_{hhh}-\frac{1+6r_h^2}{2r_h\left(1+r_h^2\right)}r_{hh}^2+\frac{r_h}{r}r_{hh}+\frac{1+r_h^2\left(3+2r_h^2\right)}{2r^2 r_h}=\frac{c_3\left(1+r_h^2\right)^{\frac{5}{2}}}{rr_h}$$

holds on the solutions of equation (3.36) where c_3 is the constant of integration. It should be remarked that a variety of boundary value problems for Willmore surfaces of revolution have been studied recently on the ground of equation (3.36), see, e.g. [5, 6, 7, 31, 32] and references therein.

Case (4). Assume $r = u$ and denote $h = w(r)$. Now, equation (3.2) reads

$$\begin{aligned}
&2r^3\left(w_r^2+1\right)^2 w_{rrrr}\\
&+4r^2(w_r^2+1)(w_r^2+1-5rw_r w_{rr})w_{rrr}\\
&+5r^3(6w_r^2-1)w_{rr}^3-15r^2 w_r(w_r^2+1)w_{rr}^2\\
&+r(w_r^2+1)^2(w_r^2-2)w_{rr}+w_r(w_r^2+1)^3(w_r^2+2)=0 \qquad (3.37)
\end{aligned}$$

and the mean and Gaussian curvatures read

$$H=\frac{1}{2r}\frac{rw_{rr}+w_r^3+w_r}{\left(1+w_r^2\right)^{3/2}}, \qquad K=\frac{1}{r}\frac{w_{rr}w_r}{\left(1+w_r^2\right)^2}. \qquad (3.38)$$

The conservation law

$$w_{rrr}-\frac{5w_r}{2\left(1+w_r^2\right)}w_{rr}^2+\frac{1}{r}w_{rr}-\frac{w_r\left(2+3w_r^2+w_r^4\right)}{2r^2}=\frac{c_4\left(1+w_r^2\right)^{\frac{5}{2}}}{r}$$

holds on the solutions of equation (3.37) where c_4 is the constant of integration. It was established in [35] that each solution of the following normal system of two ordinary differential equations

$$w_r=t, \qquad t_r=\frac{1}{r}\left(t^2+1\right)\sqrt{t^2+2\,\alpha\sqrt{t^2+1}}, \qquad \alpha\in\mathbb{R} \qquad (3.39)$$

satisfies equation (3.37) and hence determines a Willmore surface of revolution. Indeed, taking into account that the above system is equivalent to the single second-order equation

$$w_{rr} = \frac{1}{r}\left(w_r^2+1\right)\sqrt{w_r^2+2\,\alpha\sqrt{w_r^2+1}} \tag{3.40}$$

and substituting the expressions for the derivatives w_{rr}, w_{rrr} and w_{rrrr} obtained from equation (3.40) into the left-hand side of equation (3.37) one can verify that it equals zero.
Note that the mappings

$$w\to -w, \qquad w\to w+\omega, \qquad w\to w\eta, \qquad r\to r\eta, \qquad \omega,\eta\in\mathbb{R} \tag{3.41}$$

transform one solution of equation (3.37) into another, i.e., one axisymmetric Willmore surface determined in this way is mapped into another one obtained from it by reflection about the XOY-plane or scaling of the respective position vector.
Substituting equations (3.39) into equations (3.38) one obtains the following expressions for the mean and Gaussian curvatures of a surface belonging to the regarded family of axially symmetric Willmore surfaces

$$H = \frac{t+\sqrt{t^2+2\,\alpha\sqrt{tv^2+1}}}{2r\sqrt{t^2+1}}, \qquad K = \frac{t\sqrt{t^2+2\,\alpha\sqrt{t^2+1}}}{r^2\left(t^2+1\right)}$$

where $t=t(r)$ should satisfy the second equation of system (3.39).
Evidently, system (3.39) can be cast in the form

$$\begin{aligned} \frac{\mathrm{d}r}{\mathrm{d}t} &= \frac{r}{\left(t^2+1\right)\sqrt{t^2+2\alpha\sqrt{t^2+1}}} \\ \frac{\mathrm{d}w}{\mathrm{d}t} &= \frac{tr}{\left(t^2+1\right)\sqrt{t^2+2\alpha\sqrt{t^2+1}}} \end{aligned} \tag{3.42}$$

in which t plays the role of the independent variable while r and w are regarded as dependent variables. It is straightforward to present the solutions of system (3.42) by quadratures, namely

$$r(t) = a_1\exp\left[\rho(t)\right] \tag{3.43}$$

$$w(t) = a_1\int\frac{\exp\left[\rho(t)\right]t\,\mathrm{d}t}{\left(t^2+1\right)\sqrt{t^2+2\,\alpha\sqrt{t^2+1}}} + a_2 \tag{3.44}$$

where a_1 and a_2 are arbitrary real numbers and

$$\rho(t) = \int\frac{\mathrm{d}t}{\left(t^2+1\right)\sqrt{t^2+2\,\alpha\sqrt{t^2+1}}}. \tag{3.45}$$

Thus, the axially symmetric Willmore surfaces belonging to the considered

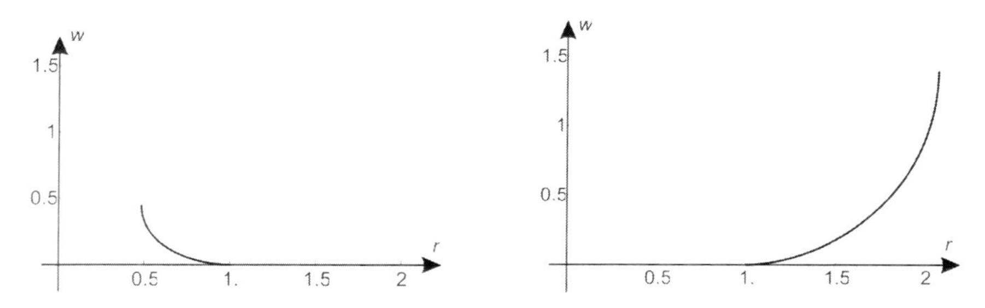

FIGURE 3.6: Parametric plots of two distinct shapes, provided by the expressions (3.46), (3.47) for $\alpha = 1$: the curve to the left corresponds to $v \in (-\infty, 0]$, and the curve to the right – to $v \in [0, \infty)$.

family are parametrized analytically in terms of the parameter t, which is nothing but the tangent of the slope angle.

It can be proved that each solution of system (3.42) is real-valued and bounded provided that $t \in (-\infty, \infty)$ if $\alpha > 0$ and $t \in (-\infty, -\varepsilon)$ or $t \in (\varepsilon, +\infty)$, where $\varepsilon = \sqrt{2\alpha^2 + 2\sqrt{\alpha^2 + \alpha^4}}$, if $\alpha < 0$. In the case $\alpha = 0$ one obtains only spheres and catenoids, see [35, p. 259].

The integrals on the right-hand sides in formulas (3.43) and (3.44) are computed numerically in the following way. First, the expression (3.45) is written in the form

$$\rho(v) = \int_a^b \frac{\mathrm{d}\tau}{(\tau^2 + 1)\sqrt{\tau^2 + 2\,\alpha\sqrt{\tau^2 + 1}}} \tag{3.46}$$

where a and b ($a \leq b$) are chosen appropriately for each case considered here and the integral is computed by the routine `NIntegrate` in `Mathematica`®. The second parametric equation (3.44) is written in the form

$$w(v) = \int_a^b \frac{\tau\, r(\tau)\,\mathrm{d}\tau}{(\tau^2 + 1)\sqrt{\tau^2 + 2\,\alpha\sqrt{\tau^2 + 1}}} \tag{3.47}$$

and in order to avoid nested `NIntegrate` routines, this integral is computed in the following manner: the interval $[a, b]$ is divided in n subintervals of equal length $\Delta\tau = (b - a)/n$ and the expression (3.47) is computed numerically using the midpoint approximation, namely

$$w(v) = \sum_{\tau=1}^{n} \Delta\tau \frac{\tilde{\tau}\, r(\tilde{\tau})}{(\tilde{\tau}^2 + 1)\sqrt{\tilde{\tau}^2 + 2\,\alpha\sqrt{\tilde{\tau}^2 + 1}}}, \qquad \tilde{\tau} = \tau + \frac{\Delta\tau}{2}$$

where for each $\tilde{\tau}$, $r(\tilde{\tau}) = \mathrm{e}^{\rho(\tilde{\tau})}$ is computed using `NIntegrate` as is mentioned above.

In the case $\alpha > 0$, the root in the right-hand side of the second equation (3.44) is real for $-\infty < v < \infty$ and formulas (3.46), (3.47) imply

$$r(0) = 1, \qquad w(0) = 0. \tag{3.48}$$

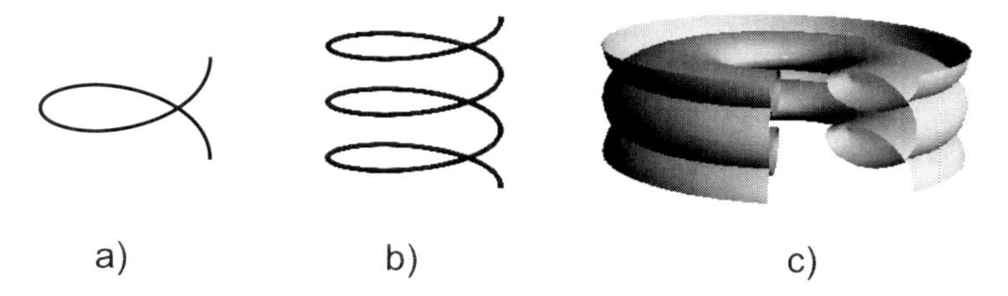

FIGURE 3.7: Parametric plots of a fragment of a nodoid a), and samples of the profile curve of this nodoid b) and its 3D shape c) for $\alpha = 1$.

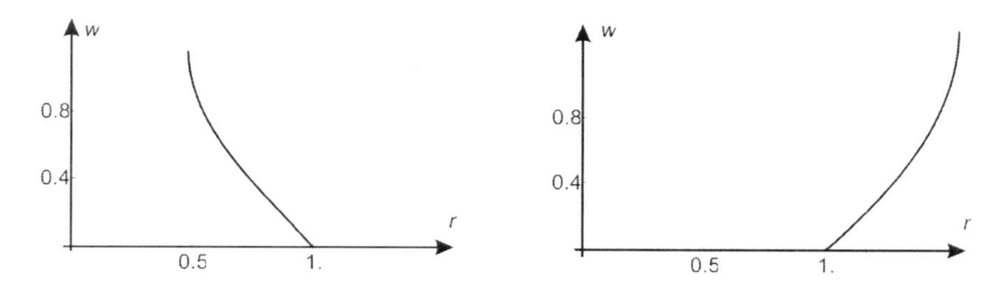

FIGURE 3.8: Parametric plots of the two shapes, provided by expressions (3.46), (3.47) for $\alpha = -0.5325$: the curve to the left corresponds to $v \in (-\infty, -\epsilon)$, and the curve to the right – to $v \in (\epsilon, \infty)$.

In fact, the expressions (3.46), (3.47) comprise two distinct fragments in (r, w)-plane, corresponding to positive or negative values of v, respectively. The boundaries of the integrals (3.46), (3.47) are chosen to be $a = 0$, $b = v$ for the positive values of v and $a = v$, $b = 0$ for the negative ones. A particular example of such fragments is shown in Fig. 3.6.

Using appropriate transformations of form (3.41) of two fragments, one can construct a variety of profile curves that are smooth solutions to system (3.39). Indeed, using the reflection $w \to -w$ and translation one obtains the fragment shown in Fig. 3.7 a). Reflecting and translating appropriately this fragment one gets the profile curve (see Fig. 3.7 b)), which gives rise to the nodoid-like surface shown in Fig. 3.7 c). In the case $\alpha < 0$ the formulas (3.48) do not hold any more. In this case, expressions (3.46), (3.47) give rise to another couple of curves in (r, w)-plane, one of which is obtained for negative values of v and the boundaries of the integrals (3.46), (3.47) are $a = v$, $b = \varepsilon$, whereas the other one is obtained for positive values of v and $a = \varepsilon$, $b = v$. A particular example of such fragments is shown in Fig. 3.8. As in the previous case, transforming these two curves by means of the mappings (3.41) one can plot a fragment of a profile curve, which corresponds to the unduloid-like surface shown in Fig. 3.9.

Finally, it is possible to combine profile curves corresponding to various values of α in order to obtain a smooth axially symmetric Willmore surface, provided

that the curvatures at the contact points are continuous. Indeed, the two fragments in Fig.3.6 can be combined into the profile, shown in Fig.3.10 a). Combining it with the fragment shown in Fig.3.9 a) one obtains the smooth profile curve Fig.3.10 b), giving raise to the surface depicted in Fig.3.10 c).
Let us now consider the Helfrich functional (3.4). Assuming $r = u$ and $h_u = h_r = \tan\psi(r)$, as in Case (1), the shape equation (3.5) reduces to the following nonlinear third-order ordinary differential equation

$$\begin{aligned} \cos^3\psi\psi_{rrr} &= 4\sin\psi\cos^2\psi\psi_{rr}\psi_r - \frac{2\cos^3\psi}{r}\psi_{rr} \\ &- \cos\psi\left(\sin^2\psi - \frac{1}{2}\cos^2\psi\right)\psi_r^3 + \frac{7\sin\psi\cos^2\psi}{2r}\psi_r^2 \\ &+ \left(\frac{\lambda}{k_c} + \frac{\mathbb{h}^2}{2} + \frac{2\mathbb{h}\sin\psi}{r} - \frac{\sin^2\psi - 2\cos^2\psi}{2r^2}\right)\cos\psi\psi_r \\ &+ \left(\frac{\lambda}{k_c} + \frac{\mathbb{h}^2}{2} - \frac{\sin^2\psi + 2\cos^2\psi}{2r^2}\right)\frac{\sin\psi}{r} - \frac{p}{k_c} \end{aligned} \tag{3.49}$$

derived by Hu and Ou-Yang in [15].
In 1995, Naito *et al.* [21] discovered (see also [24]) that the shape equation for axisymmetric fluid membranes (3.49) has the following class of exact solutions

$$\sin\psi = ar + b + dr^{-1} \tag{3.50}$$

provided that a, b and d are real constants which meet the conditions

$$\frac{p}{k_c} - 2a^2\mathbb{h} - 2a\left(\frac{\mathbb{h}^2}{2} + \frac{\lambda}{k_c}\right) = 0 \tag{3.51}$$

$$b\left(2a\mathbb{h} + \frac{\mathbb{h}^2}{2} + \frac{\lambda}{k_c}\right) = 0 \tag{3.52}$$

$$b\left(b^2 - 4ad - 4\mathbb{h}d - 2\right) = 0 \tag{3.53}$$

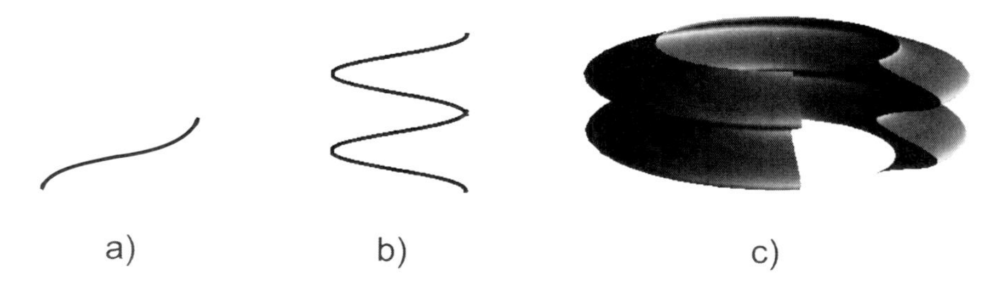

FIGURE 3.9: Parametric plots of a fragment of an unduloid a), and samples of the profile curve of this unduloid b) and its 3D shape c) for $\alpha = -1$.

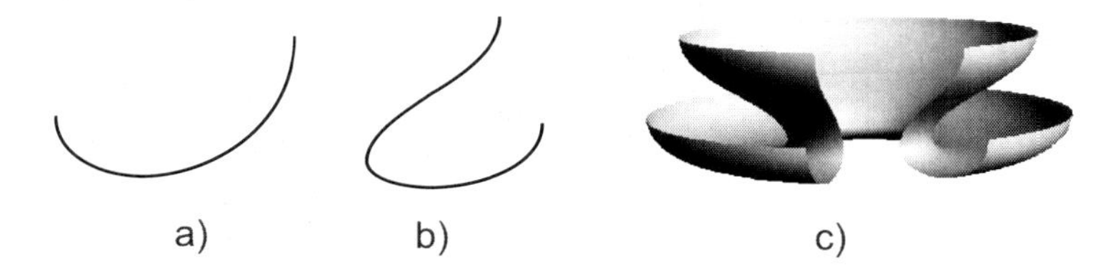

FIGURE 3.10: Parametric plots of a fragment of a nodoid a), a profile curve b) combining this fragment and the one in Fig. 3.9 a) and the corresponding 3D shape c).

and

$$d\left(b^2 - 4ad - 2\mathbb{h}d\right) = 0. \tag{3.54}$$

Six types of solutions (3.50) to equation (3.49) can be distinguished on the ground of conditions (3.51) – (3.54) depending on the values of the constants $\mathbb{h}$, λ and p.

Case A. If $\mathbb{h} = 0$, $\lambda = 0$, $p = 0$, then the solutions to equation (3.49) of the form (3.50) are $\sin\psi = ar$, $\sin\psi = ar \pm \sqrt{2}$ and $\sin\psi = dr^{-1}$, the respective surfaces being spheres, Clifford tori and catenoids.

Case B. If $\mathbb{h} = 0$, $\lambda \neq 0$, $p = 0$, then the solutions of the considered type reduce to $\sin\psi = dr^{-1}$ (catenoids).

Case C. If $\mathbb{h} = 0$, $\lambda \neq 0$, $p \neq 0$ and $p = 2a\lambda$, then only one branch of the regarded solutions remains, namely $\sin\psi = ar$ (spheres).

Case D. If $\mathbb{h} \neq 0$, $\lambda = 0$, $p = 0$, then one arrives at the whole family of Delaunay surfaces corresponding to the solutions of the form

$$\sin\psi = -\frac{1}{2}\mathbb{h}r + \frac{d}{r}\cdot \tag{3.55}$$

Case E. If $\mathbb{h} \neq 0$, $\lambda \neq 0$, $p = 0$ and

$$\frac{\lambda}{k_c} = -\frac{1}{2}\mathbb{h}\left(2a + \mathbb{h}\right)$$

one gets only solutions of the form $\sin\psi = ar$ (spheres).

Case F. If $\mathbb{h} \neq 0$, $\lambda \neq 0$, $p \neq 0$, then four different types of solutions of form (3.50) to equation (3.49) are encountered: a) $\sin\psi = ar$ (spheres) if

$$\frac{p}{k_c} = 2a\left(\frac{\lambda}{k_c} + a\mathbb{h} + \frac{\mathbb{h}^2}{2}\right)$$

b) $\sin\psi = ar \pm \sqrt{2}$ (Clifford tori) if

$$\frac{p}{k_c} = -2a^2\mathbb{h}, \qquad \frac{\lambda}{k_c} = -\frac{1}{2}\mathbb{h}\left(4a + \mathbb{h}\right)$$

c) solutions of the form (3.55) (Delaunay surfaces) if

$$p + \mathbb{h}\lambda = 0$$

d) solutions of the form

$$\sin\psi = -\frac{1}{4}\mathrm{Ih}\left(b^2+2\right)r + b - \frac{1}{\mathrm{Ih}r} \tag{3.56}$$

which take place provided that

$$\frac{p}{k_c} = -\frac{1}{8}\mathrm{Ih}^3\left(b^2+2\right)^2, \qquad \frac{\lambda}{k_c} = \frac{1}{2}\mathrm{Ih}^2\left(b^2+1\right).$$

Below, following [8], we derive the parametric equations of the surfaces corresponding to the solutions of form (3.56) to equation (3.49).
First, it is clear that the variable r must be strictly positive or negative, otherwise the right-hand side of equation (3.50) is both undefined and its absolute value is greater than one, which is in contradiction with the sin-function appearing in the left-hand side of this relation.
For the regarded class of solutions (3.56) the relation

$$\frac{\mathrm{d}h}{\mathrm{d}r} = \tan\psi$$

implies

$$\left(\frac{\mathrm{d}h}{\mathrm{d}r}\right)^2 = \frac{\left(b - \frac{1}{\mathrm{Ih}r} - \frac{1}{4}\mathrm{Ih}\left(b^2+2\right)r\right)^2}{1 - \left(b - \frac{1}{\mathrm{Ih}r} - \frac{1}{4}\mathrm{Ih}\left(b^2+2\right)r\right)^2}. \tag{3.57}$$

In terms of an appropriate variable u, relation (3.57) may be written in the form

$$\left(\frac{\mathrm{d}r}{\mathrm{d}u}\right)^2 = -\frac{1}{\bar{a}^2}Q_1(r)Q_2(r) \tag{3.58}$$

$$\left(\frac{\mathrm{d}h}{\mathrm{d}u}\right)^2 = \frac{1}{4\bar{a}^2}\left(Q_1(r)+Q_2(r)\right)^2 \tag{3.59}$$

where

$$\bar{a} = -\frac{4}{\mathrm{Ih}\left(2+b^2\right)^{3/4}}$$

$$Q_1(r) = r^2 - \frac{4(b+1)}{\mathrm{Ih}\left(b^2+2\right)}r + \frac{4}{\mathrm{Ih}^2\left(b^2+2\right)} \tag{3.60}$$

$$Q_2(r) = r^2 - \frac{4(b-1)}{\mathrm{Ih}\left(b^2+2\right)} + \frac{4}{\mathrm{Ih}^2\left(b^2+2\right)}. \tag{3.61}$$

It should be noticed that the roots of the polynomial $Q(r) = Q_1(r)Q_2(r)$ read

$$\alpha = \frac{2\,\mathrm{sign}\,(b)}{\mathrm{Ih}\sqrt{b^2+2}}\frac{\chi-1}{\chi+1}, \qquad \beta = \frac{2\,\mathrm{sign}\,(b)}{\mathrm{Ih}\sqrt{b^2+2}}\frac{\chi+1}{\chi-1} \tag{3.62}$$

$$\gamma = \frac{4b}{\mathrm{Ih}\left(b^2+2\right)} - \frac{\alpha+\beta}{2} + \mathrm{i}\frac{2\sqrt{2|b|+1}}{\mathrm{Ih}\left(b^2+2\right)}$$

$$\delta = \frac{4b}{\mathbb{h}\,(b^2+2)} - \frac{\alpha+\beta}{2} - \mathrm{i}\frac{2\sqrt{2|b|+1}}{\mathbb{h}\,(b^2+2)}$$

where

$$\chi = \sqrt{\frac{1+|b|+\sqrt{2+b^2}}{1+|b|-\sqrt{2+b^2}}}. \tag{3.63}$$

Hence, equation (3.58) has real-valued solutions if and only if at least two of these roots are real and different. Evidently, the roots γ and δ can not be real, but α and β are real provided that $|b| > 1/2$ as follows by the relations (3.62) and (3.63).

Now, using the standard procedure for handling elliptic integrals (see [37, Section 20.6]), one can express the solution $r(u)$ of equation (3.58) in the form

$$r(u) = \frac{2\,\mathrm{sign}\,(b)}{\mathbb{h}\sqrt{b^2+2}}\left(1 - \frac{2\chi}{\chi+\mathrm{cn}(u,k)}\right) \tag{3.64}$$

where

$$k = \sqrt{\frac{1}{2} - \frac{3}{4\sqrt{2+b^2}}}.$$

Consequently, using expressions (3.60) and (3.61), one can write down the solution $z(t)$ of equation (3.59) in the form

$$h(u) = \frac{1}{\bar{a}}\int\left(r^2(u) - \frac{4\,b\,r(u)}{\mathbb{h}\,(b^2+2)} + \frac{4}{\mathbb{h}^2\,(b^2+2)}\right)\mathrm{d}u$$

that is

$$h(u) = \bar{a}\left(E(\mathrm{am}(u,k),k) - \frac{\mathrm{sn}(u,k)\,\mathrm{dn}(u,k)}{h+\mathrm{cn}(u,k)} - \frac{u}{2}\right). \tag{3.65}$$

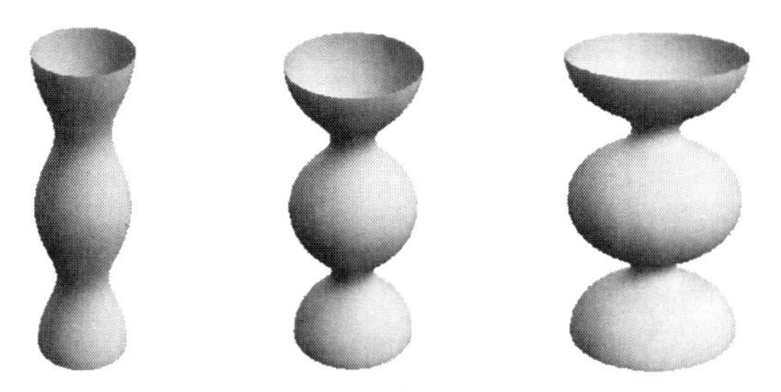

FIGURE 3.11: Unduloid-like Helfrich surfaces obtained using the parametric equations (3.64) and (3.65).

Thus, for each couple of values of the parameters $\mathbb{h}$ and b, (3.64) and (3.65) are the parametric equations of the contour of an axially symmetric surface corresponding to the respective solution of the membrane shape equation (3.49) of form (3.56). Examples of such surfaces are given in Fig. 3.11.

3.4 Dupin Cyclides

There exist numerous definitions of Dupin cyclides and we refer to Krivoshapko & Ivanov [18] for the most basic notions, illustrations and representative list of references. Here we will only mention that they are algebraic surfaces of fourth order which can be represented implicitly by the equation

$$\left(x^2+y^2+z^2+b^2-\mu^2\right)^2-4(cz+a\mu)^2-4b^2x^2=0 \tag{3.66}$$

where a, b, μ, c are positive real parameters fulfilling the conditions $c > a$, $b = \sqrt{c^2-a^2}$ and $c > \mu > a$. The uniformization of the quartic (3.66) is provided by the real parameters $u, v \in [0, 2\pi]$ through the coordinate functions x, y, z, i.e.,

$$\mathbf{x}=\left(\frac{b(c-\mu\cos v)\sin u}{a\cos u\cos v-c},\frac{b(\mu-a\cos u)\sin v}{a\cos u\cos v-c},\frac{b^2\cos u+\mu(a-c\cos u\cos v)}{a\cos u\cos v-c}\right)^T$$

which define the smooth surface $\mathcal{S}$. By making use of this parameterization it is a matter of lengthy calculations to find the coefficients E, F, G of the first fundamental form

$$\begin{aligned}
E &= \mathbf{x}_u\cdot\mathbf{x}_u=\frac{\left(c^2-a^2\right)(\mu\cos v-c)^2}{(a\cos u\cos v-c)^2}\\
F &= \mathbf{x}_u\cdot\mathbf{x}_v=0\\
G &= \mathbf{x}_v\cdot\mathbf{x}_v=\frac{\left(c^2-a^2\right)(a\cos u-\mu)^2}{(a\cos u\cos v-c)^2}.
\end{aligned}$$

Respectively, the coefficients L, M, N of the second fundamental form are

$$\begin{aligned}
L &= \mathbf{x}_{uu}\cdot\mathbf{n}=\frac{\left(c^2-a^2\right)(\mu\cos v-c)\cos v}{(a\cos u\cos v-c)^2}\\
M &= \mathbf{x}_{uv}\cdot\mathbf{n}=0\\
N &= \mathbf{x}_{vv}\cdot\mathbf{n}=\frac{\left(c^2-a^2\right)(a\cos u-\mu)}{(a\cos u\cos v-c)^2}
\end{aligned} \tag{3.67}$$

which are evaluated with the help of the unit normal vector to the surface $\mathcal{S}$

$$\mathbf{n}=\frac{\mathbf{x}_u\times\mathbf{x}_v}{|\mathbf{x}_u\times\mathbf{x}_v|}=\left(\frac{b\cos v\sin u}{a\cos u\cos v-c},\frac{b\sin v}{c-a\cos u\cos v},\frac{a-c\cos u\cos v}{c-a\cos u\cos v}\right)^T.$$

Having at our disposition the first and the second fundamental forms, we can evaluate the surface invariants like the mean H and Gaussian K curvatures, or the expression for the surface Laplacian $\Delta_{\mathcal{S}}$. Actually, we have

$$H = \frac{c+(a\cos u-2\mu)\cos v}{2(a\cos u-\mu)(\mu\cos v-c)}, \qquad K=\frac{\cos v}{(\mu-a\cos u)(\mu\cos v-c)}$$

$$\begin{aligned}\Delta_{\mathcal{S}} &= \frac{(c-a\cos u\cos v)^2}{(c^2-a^2)(\mu-a\cos u)^2(c-\mu\cos v)^2}\Bigg((\mu-a\cos u)((\mu-a\cos u)\frac{\partial^2}{\partial u^2}\\ &\quad +a\sin u\frac{\partial}{\partial u}) + (c-\mu\cos v)^2\frac{\partial^2}{\partial v^2} - \mu(\mu\cos v - c)\sin v\frac{\partial}{\partial v}\Bigg).\end{aligned} \tag{3.68}$$

These are all ingredients needed for writing down equation (3.2) and the result of doing this is a polynomial whose coefficients in front of the various powers of $\cos^m u\cos^n v$ for all possible combinations of $m, n = 0, 1, 2, 3$ are functions of the geometrical parameters a, c and μ. All these coefficients miraculously vanish provided that

$$\mu = \sqrt{\frac{a^2+c^2}{2}} \tag{3.69}$$

is fulfilled and this means that surfaces described by the equations (3.67) in which the parameter μ satisfies the above condition are Willmore surfaces. At this point it is reasonable to ask if there are some other constraints on the parameters which will provide the absolute minima of the functional (3.1). In order to answer this question we have to evaluate (3.1) and for that purpose we need the infinitesimal surface area element $\mathrm{d}\mathcal{A}$ associated with the surface metric, i.e.,

$$\mathrm{d}\mathcal{A} = \sqrt{EG-F^2}\mathrm{d}u\wedge\mathrm{d}v = \frac{(a^2-c^2)(\mu-a\cos u)(c-\mu\cos v)}{(c-a\cos u\cos v)^2}\mathrm{d}u\wedge\mathrm{d}v. \tag{3.70}$$

Taking into account (3.68), (3.70) and performing the necessary integrations we end up with the following expression

$$W = 2\frac{(\sqrt{2}\sqrt{t^2+1}-t-1)\sqrt{3t^2+2\sqrt{2}\sqrt{t^2+1}t+1}}{(1-t)\sqrt{t^2-2\sqrt{2}\sqrt{t^2+1}+3}}\pi^2$$

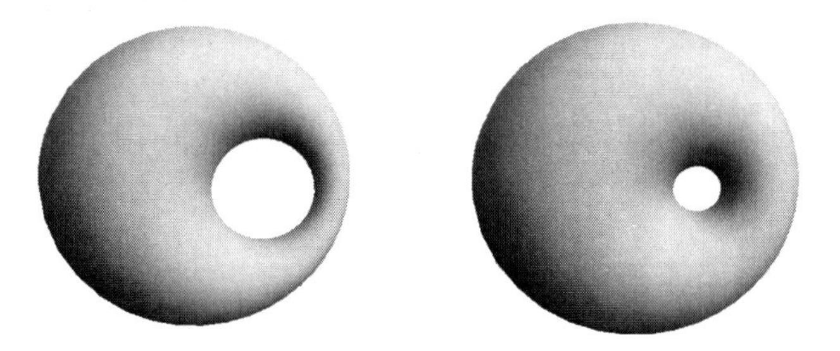

FIGURE 3.12: Dupin cyclides pictured with parameters $a = 1, c = 3$, $\mu = 1.5$ (left) and μ given by formula (3.69) (right).

in which the parameter t is introduced in place of a/c. Having in mind the conventions fixed at the beginning it is obvious that $t \in (0, 1)$. Also, it is a matter of simple calculations to establish that the fraction in the above formula is a constant in the defining interval. Strange or not its value there is just one! All this means that the Dupin cyclides like the Clifford tori provide the absolute minima of the Willmore functional equal to $2\pi^2$. Actually, this fact is not entirely surprising as it is well known that the cyclides are images of tori under inversions and that the latter are conformal mappings which preserve the Willmore functional.
It is noteworthy that in [27] the Dupin cyclide was found to describe deformed shapes of smectic A liquid crystal lamellae.

3.5 Willmore Surfaces Obtained by Inversions

As it was already mentioned in the introduction, the Willmore functional is invariant under the 10-parameter group of conformal transformations in $\mathbb{R}^3$. Among them are the inversions, which constitute a 3-parameter subgroup G_{SCT} of special conformal transformations defined as follows

$$\mathbf{x}' = \frac{\mathbf{x} - \mathbf{a}\mathbf{x}^2}{1 - 2\mathbf{a} \cdot \mathbf{x} + \mathbf{a}^2\mathbf{x}^2}$$

where $\mathbf{a} = (a_1, a_2, a_3)$ is a vector.
In [16, 17, 28], conformal transformations of this special type were used to study the morphology of lipid bilayer vesicles and the stability of their equilibrium shapes. Inverted catenoids (see Fig. 3.13) were used in [3, 4] to obtain a one-parameter family of exact solutions of the shape equation (3.5) displaying a discocyte to stomatocyte transition in axially symmetric vesicles with polar tethers as well as to analyze the external forces required to pull the poles together and the stresses in the inverted shapes.

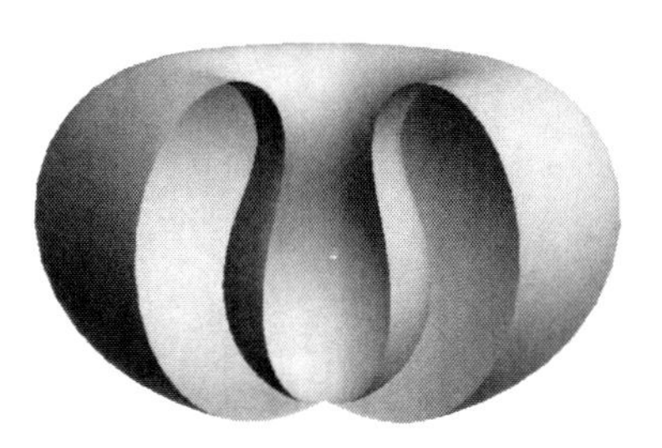

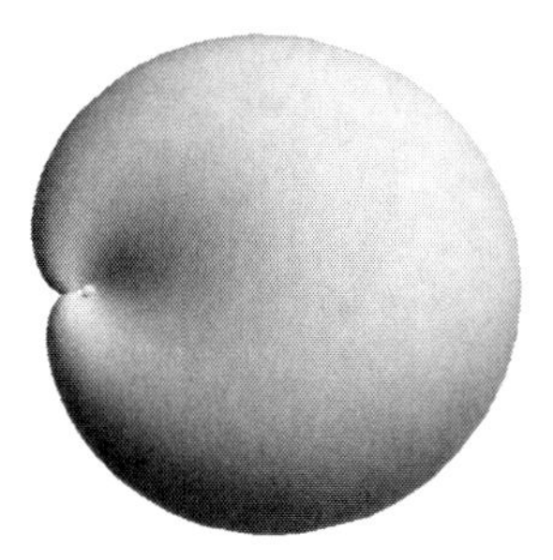

FIGURE 3.13: Inverted catenoid (left) and Enneper surface (right).

It is noteworthy that the Willmore energy of surfaces depicted in Figs. 3.13 is equal to zero.

Appendix. Elliptical Integrals and Jacobian Elliptic Functions

The elliptic integrals and functions are mathematical objects, which nowadays are often omitted in the mathematics curricula of the universities. One quite trivial explanation is the presence of plenty of efficient computational programs that can be implemented on modern computers.

While the standard integration techniques allow us to obtain explicit expressions (in terms of trigonometric, exponential and logarithmic functions) for every integral in the form

$$\int \mathcal{R}(x, \sqrt{P(x)})\mathrm{d}x$$

where $\mathcal{R}(x, \sqrt{P(x)})$ is rational function, and $P(x)$ is linear or quadratic polynomial, we have to widen our vocabulary of "elementary" functions if we want to work with polynomials of higher degree. In particular when $P(x)$ is a polynomial of third or fourth degree the corresponding function is called **elliptic**. Of course, when someone teaches calculus, he must stop somewhere and it is reasonably simple to stay loyal to well known linear and quadratic functions, while using numerical methods to calculate integrals of the third and fourth degree. The possibilities of easy-to-use computer systems for symbolic manipulation of the type of *Maple*® and `Mathematica`® make this course of action even more understandable.

The main point in this Appendix is that elliptic functions provide effective means for the description of geometric objects. The second is that the above mentioned computer programs through their built-in tools for calculation and visualization are in fact real motivation for teaching and using elliptic functions.

In this Appendix we will consider a few examples in order to prove that elliptic integrals and functions are necessary to get interesting geometric and mechanical information more than that given from direct numerical calculations.

The history of the development of elliptic functions can be followed in [29]. Clear statement of their properties and applications can be found in the books [1, 11, 13, 19]. Today's approach to the problem from the viewpoint of the dynamical systems is given by [20].

The easiest way to understand the elliptic functions is to consider them as analogous to the ordinary trigonometric functions. From the calculus we know that

$$\arcsin(x) = \int_0^x \frac{du}{\sqrt{1-u^2}}.$$

Of course, if $x = \sin(t)$, $-\pi/2 \le t \le \pi/2$, we will have

$$t = \arcsin(\sin(t)) = \int_0^{\sin(t)} \frac{du}{\sqrt{1-u^2}}. \tag{3.71}$$

In this case we can consider $\sin(t)$ as the inverse function of the integral (3.71). The real understanding of trigonometric functions includes knowledge of their graphics, their connection with other trigonometric functions as $\sin^2(\theta) + \cos^2(\theta) = 1$ and of course the fundamental geometric and physical parameters in which they are included (i.e., circumferences and periodical movements). We will follow this example for elliptic functions too.
Let us begin by fixing some k, $0 \le k \le 1$ which from now will be called **elliptic modulus** and introduce the following

Definition 3.7. *The Jacobi sine function* $\mathrm{sn}(u,k)$ *is the inverse function of the following integral*

$$u = \int_0^{\mathrm{sn}(u,k)} \frac{dt}{\sqrt{1-t^2}\sqrt{1-k^2t^2}}. \tag{3.72}$$

More generally, we will call

$$F(z,k) = \int_0^z \frac{dt}{\sqrt{1-t^2}\sqrt{1-k^2t^2}} \tag{3.73}$$

the elliptic integral of the first kind. The elliptic integrals of the second and third kind are defined by the equations

$$\begin{aligned} E(z,k) &= \int_0^z \frac{\sqrt{1-k^2t^2}}{\sqrt{1-t^2}}\,dt \\ \Pi(n,z,k) &= \int_0^z \frac{dt}{(1+nt^2)\sqrt{(1-t^2)(1-k^2t)}}. \end{aligned}$$

When the argument z in $F(z,k), E(z,k)$ and $\Pi(n,z,k)$ is equal to one, these integrals are denoted respectively as $K(k), E(k)$ and $\Pi(n,k)$ and called complete elliptic integrals of the first, second and third kind, respectively.
If we put $t = \sin\phi$ the above integrals are transformed respectively in

$$\begin{aligned} F(\phi,k) &= \int_0^\phi \frac{d\phi}{\sqrt{1-k^2\sin^2\phi}} \\ E(\phi,k) &= \int_0^\phi \sqrt{1-k^2\sin^2\phi}\,d\phi \end{aligned}$$

$$\Pi(n,\phi,k) \quad = \quad \int_0^{\phi} \frac{\mathrm{d}\phi}{(1+n\sin^2\phi)\sqrt{1-k^2\sin^2\phi}}.$$

Let us note that when $k \equiv 1$, $E(\phi,1) = \sin\phi$ and therefore one can consider $E(\phi,k)$ as a generalization of the function $\sin\phi$.

The Jacobi cosine function $\mathrm{cn}(u,k)$ can be defined in terms of $\mathrm{sn}(u,k)$ by means of the identity

$$\mathrm{sn}^2(u,k) + \mathrm{cn}^2(u,k) = 1. \tag{3.74}$$

The third Jacobi elliptic function $\mathrm{dn}(u,k)$ is defined by the equation

$$\mathrm{dn}^2(u,k) + k^2\,\mathrm{sn}^2(u,k) = 1. \tag{3.75}$$

The integral definition of $\mathrm{sn}(u,k)$ makes it clear that $\mathrm{sn}(u,0) = \sin(u)$. Of course, $\mathrm{cn}(u,0) = \cos(u)$ as well.
Let us remark that Jacobi defined a function $\mathrm{am}(z,k)$ by means of the equation

$$\mathrm{am}(z,k) = \int_0^z \mathrm{dn}(s,k)\mathrm{d}s.$$

In terms of this function, which is nowadays referred to as the Jacobi amplitude function, see e.g. [19], one has the useful representation

$$\mathrm{sn}(z,k) = \sin(\mathrm{am}(z,k)), \qquad \mathrm{cn}(z,k) = \cos(\mathrm{am}(z,k)).$$

Besides sn, cn and dn there are another nine functions, which are widely used and their definitions are given below

$$\begin{aligned}
\mathrm{ns} &= \frac{1}{\mathrm{sn}}, & \mathrm{nc} &= \frac{1}{\mathrm{cn}}, & \mathrm{nd} &= \frac{1}{\mathrm{dn}} \\
\mathrm{sc} &= \frac{\mathrm{sn}}{\mathrm{cn}}, & \mathrm{cd} &= \frac{\mathrm{cn}}{\mathrm{dn}}, & \mathrm{ds} &= \frac{\mathrm{dn}}{\mathrm{sn}} \\
\mathrm{cs} &= \frac{\mathrm{cn}}{\mathrm{sn}}, & \mathrm{dc} &= \frac{\mathrm{dn}}{\mathrm{cn}}, & \mathrm{sd} &= \frac{\mathrm{sn}}{\mathrm{dn}}.
\end{aligned}$$

The derivatives of the elliptic functions can be found directly by their definitions (or vice versa like in [20] where the elliptical functions are defined by their derivatives). For instance, the derivative of $\mathrm{sn}(u,k)$ may be computed as follows. In (3.73) suppose that $z = z(u)$. Then

$$\frac{\mathrm{d}F}{\mathrm{d}u} = \frac{\mathrm{d}F}{\mathrm{d}z}\frac{\mathrm{d}z}{\mathrm{d}u} = \frac{1}{\sqrt{1-z^2}\sqrt{1-k^2z^2}}\frac{\mathrm{d}z}{\mathrm{d}u}.$$

But from (3.72) and (3.73) we know that for $z = \mathrm{sn}(u,k)$, we have $F(z,k) = u$. So, replacing z by $\mathrm{sn}(u,k)$ and using $\mathrm{d}u/\mathrm{d}u = 1$, we obtain

$$1 \quad = \quad \frac{1}{\sqrt{1-\mathrm{sn}(u,k)^2}\sqrt{1-k^2\mathrm{sn}(u,k)^2}}\frac{\mathrm{d}\,\mathrm{sn}(u,k)}{\mathrm{d}u}$$

$$\begin{aligned}\frac{\mathrm{d}\,\mathrm{sn}(u,k)}{\mathrm{d}u} &= \sqrt{1-\mathrm{sn}(u,k)^2}\sqrt{1-k^2\mathrm{sn}(u,k)^2} \\ \frac{\mathrm{d}\,\mathrm{sn}(u,k)}{\mathrm{d}u} &= \mathrm{cn}(u,k)\,\mathrm{dn}(u,k).\end{aligned} \tag{3.76}$$

After differentiation of (3.74) with respect of u and taking into account (3.76) we obtain

$$\frac{\mathrm{d}\,\mathrm{cn}(u,k)}{\mathrm{d}u} = -\mathrm{sn}(u,k)\,\mathrm{dn}(u,k)$$

Finally, after differentiating (3.75) and using once more (3.76), we have

$$\frac{\mathrm{d}\,\mathrm{dn}(u,k)}{\mathrm{d}u} = -k^2\mathrm{sn}(u,k)\,\mathrm{cn}(u,k).$$

The algebraic computer programs *Maple*® or `Mathematica`® have embeded modules for working with elliptic functions, so they can be easily drawn. Graphs of the elliptic sin function sn, cos function cn and function dn are shown in Fig. 3.14. We can see that $\mathrm{sn}(u,k)$ and $\mathrm{cn}(u,k)$ are periodic. We can define their period referring to the definitions above (see (3.72))

$$K(k) = \int_0^1 \frac{\mathrm{d}t}{\sqrt{1-t^2}\sqrt{1-k^2t^2}}.$$

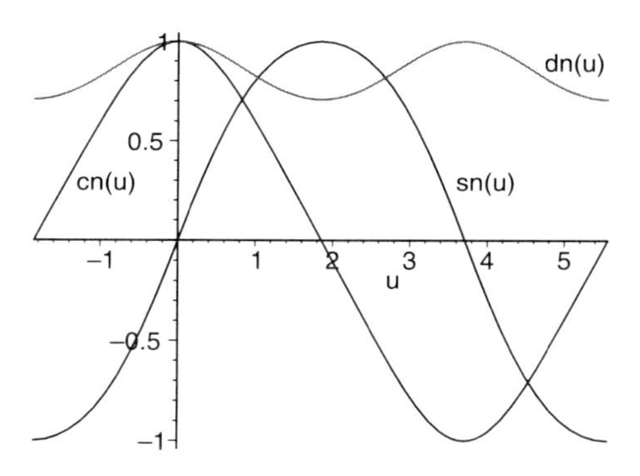

FIGURE 3.14: Graphs of the elliptic sin function $\mathrm{sn}(u,k)$, elliptic cos function $\mathrm{cn}(u,k)$ and the function $\mathrm{dn}(u,k)$ drawn with $k=\frac{1}{\sqrt{2}}$.

We can see that $\mathrm{sn}(K(k),k)=1$. Obviously, from the graph we are convinced also that $K(k)$ is $1/4$ from $\mathrm{sn}(u,k)$ period and that the period of $\mathrm{dn}(u,u)$ is $2K(u)$. Of course, this can be checked analytically (see Whittaker [37]), but this argument satisfies our objectives too.

Using the computer program, we can look for numerical solution, from which we can find $K(k)$, i.e., solve the equation $\mathrm{sn}(u,k)=1$. Note the identity $\mathrm{sn}^2(u,k)+\mathrm{cn}^2(u,k)=1$. Suppose that $\mathrm{cn}(u,k)$ have the same period $\mathrm{sn}(u,k)$ and therefore $\mathrm{cn}(K(k),k)=0$.

Now we have an idea for the algebraic and graphic properties of the elliptic functions. In order to "complement" our understanding, let us look at two simple examples - physical and geometrical - where they appear.

Example 3.8. (Pendulum) *Let the angle of a pendulum swing be denoted by x. Then it is straightforward to derive that the equation of motion is : $\ddot{x}+(g/l)\sin(x) = 0$, where g is the acceleration due to gravity and l is the length of the pendulum. If we take such units that give $g/l = 1$ the pendulum equation becomes $\ddot{x} + \sin(x) = 0$. Then, we can multiply by $\dot{x}$ to obtain*

$$\begin{aligned} \dot{x}(\ddot{x} + \sin(x)) &= 0 \\ \dot{x}\ddot{x} + 2\sin(x)\frac{\dot{x}}{2} &= 0 \\ \dot{x}\ddot{x} + 4\sin\left(\frac{x}{2}\right)\cos\left(\frac{x}{2}\right)\frac{\dot{x}}{2} &= 0 \end{aligned}$$

and by integrating the last equation to end up with

$$\frac{1}{2}\dot{x}^2 + 2\sin^2\left(\frac{x}{2}\right) = c.$$

Note also that because of the identity $2\sin^2(x/2) = 1-\cos(x)$, the last equation expresses the conservation of the energy of the motion of a particle with a unit mass. Now let $z = \sin(x/2)$ and therefore $2\dot{z} = \cos(x/2)\dot{x} = \sqrt{1-z^2}\,\dot{x}$. Then

$$\begin{aligned} 4\dot{z}^2 &= (1-z^2)\dot{x}^2 = \dot{x}^2 - \sin^2\left(\frac{x}{2}\right)\dot{x}^2 = \dot{x}^2\cos^2\left(\frac{x}{2}\right) \\ \dot{z}^2 &= \frac{1}{4}\dot{x}^2\cos^2\left(\frac{x}{2}\right). \end{aligned}$$

By the first part of the calculation, we have

$$\frac{1}{4}\dot{x}^2 = \frac{1}{2}c - \sin^2\left(\frac{x}{2}\right) = \frac{1}{2}c - z^2 \qquad \textit{and} \qquad \cos^2\left(\frac{x}{2}\right) = 1 - z^2.$$

Hence, $\dot{z}^2 = (A - z^2)(1 - z^2)$, where $A = c/2$. Taking a square root and separating the variables gives

$$t = \int_0^z \frac{\mathrm{d}z}{\sqrt{(A-z^2)(1-z^2)}} = \int_0^{\sqrt{A}u} \frac{\mathrm{d}u}{\sqrt{(1-u^2)(1-Au^2)}} = F(\sqrt{A}u, \sqrt{A}).$$

That is, we see that the elliptic integrals appear even in most standard mechanical situations.

Example 3.9. (Ellipse) *Let us parameterize the ellipse by the polar angle, which we will denote by t, i.e., $\alpha(t) = (x(t), z(t)) = (a\sin(t), c\cos(t))$, where $0 \leq t \leq 2\pi$, and $a \geq c$. Then the arclength integral is*

$$L = \int_0^{2\pi} \sqrt{\dot{x}^2 + \dot{z}^2}\,\mathrm{d}t = 4\int_0^{\pi/2} \sqrt{a^2\cos^2(t) + c^2\sin^2(t)}\,\mathrm{d}t$$

$$= \quad 4a \int_0^{2\pi} \sqrt{1 - \varepsilon^2 \sin^2(t)}\, \mathrm{d}t$$

in which $\varepsilon = \sqrt{a^2 - c^2}/a$ *is the eccentricity of the ellipse. If we substitute* $\sin(t) = u$*, then* $\mathrm{d}t = \mathrm{d}u/\sqrt{1-u^2}$ *and in this way we obtain*

$$L = 4a \int_0^1 \frac{\sqrt{1-\varepsilon^2 u^2}}{\sqrt{1-u^2}}\, \mathrm{d}u = 4a\, E(\varepsilon).$$

So we have been convinced once more again that the elliptic integrals present themselves even in the most natural geometric problems.
All Jacobian elliptic function have integrals, given by the formulas below, that can be verified by direct differentiations

$$\int \mathrm{sn}u\, \mathrm{d}u = \frac{1}{k} \ln(\mathrm{dn}u - k\mathrm{cn}u), \qquad \int \mathrm{cn}u\, \mathrm{d}u = \frac{1}{k} \arcsin(k\mathrm{sn}u)$$

$$\int \mathrm{dn}u\, \mathrm{d}u = \arcsin(\mathrm{sn}u), \qquad \int \mathrm{ns}u\, \mathrm{d}u = -\ln(\mathrm{ds}u + \mathrm{cs}u)$$

$$\int \mathrm{nc}u\, \mathrm{d}u = \frac{1}{\tilde{k}} \ln(\mathrm{dc}u + \tilde{k}\mathrm{sc}u), \qquad \int \mathrm{nd}u\, \mathrm{d}u = \frac{1}{\tilde{k}} \arcsin(\mathrm{cd}u)$$

$$\int \mathrm{sc}u\, \mathrm{d}u = \frac{1}{\tilde{k}} \ln(\mathrm{dc}u + \tilde{k}\mathrm{nc}u), \qquad \int \mathrm{cd}u\, \mathrm{d}u = \frac{1}{k} \ln(\mathrm{nc}u + \mathrm{sc}u)$$

$$\int \mathrm{ds}u\, \mathrm{d}u = \ln(\mathrm{ns}u - \mathrm{cs}u), \qquad \int \mathrm{cs}u\, \mathrm{d}u = -\ln(\mathrm{ns}u + \mathrm{ds}u)$$

$$\int \mathrm{dc}u\, \mathrm{d}u = \ln(\mathrm{nc}u + \mathrm{sc}u), \qquad \int \mathrm{sd}u\, \mathrm{d}u = -\frac{1}{k\tilde{k}} \arcsin(k\mathrm{cd}u).$$

Bibliography

[1] F. Bowman. *Introduction to elliptic functions with applications.* English Universities Press, London, 1953.

[2] R. Capovilla, C. Chryssomalakos, and J. Guven. Elastica hypoarealis. *The European Physical Journal B - Condensed Matter*, 29(2):163–166, Sep 2002.

[3] P. Castro-Villarreal and J. Guven. Axially symmetric membranes with polar tethers. *Journal of Physics A: Mathematical and Theoretical*, 40(16):4273, 2007.

[4] P. Castro-Villarreal and J. Guven. Inverted catenoid as a fluid membrane with two points pulled together. *Phis. Rev. E*, 76(1):011922, 2007.

[5] A. Dall'Acqua, K. Deckelnick, and H.-C. Grunau. Classical solutions to the Dirichlet problem for Willmore surfaces of revolution. *Advances in Calculus of Variations*, 1(4):379–397, 2008.

[6] A. Dall'Acqua, K. Deckelnick, and G. Wheeler. Unstable Willmore surfaces of revolution subject to natural boundary conditions. *Calculus of Variations and Partial Differential Equations*, 48(3-4):293–313, 2013.

[7] A. Dall'Acqua, S. Fröhlich, H.-C. Grunau, and F. Schieweck. Symmetric Willmore surfaces of revolution satisfying arbitrary Dirichlet boundary data. *Advances in Calculus of Variations*, 4(1):1–81, 2011.

[8] P. A. Djondjorov, M. T. Hadzilazova, I. M. Mladenov, and V. M. Vassilev. Beyond Delaunay sufaces. *J. Geom. Symm. Phys.*, 16:1–12, 2010.

[9] D. J. Dunstan. Continuum modelling of nanotubes: Collapse under pressure. In *Structure and Multiscale Mechanics of Carbon Nanomaterials*, pages 181–190. Springer, 2016.

[10] S. Germain. *Recherches sur la théorie des surfaces élastiques*. Veuve Courcier, Paris, 1821.

[11] A. G. Greenhill. *The applications of elliptic functions*. Dover, New York, 1959.

[12] H.-C. Grunau. Nonlinear questions in clamped plate models. *Milan Journal of Mathematics*, 77:171–204, 2009.

[13] H. Hancock. *Elliptic integrals*. Dover, New York, 1958.

[14] W. Helfrich. Elastic properties of lipid bilayers: theory and possible experiments. *Zeitschrift für Naturforschung C*, 28:693–703, 1973.

[15] J.-G. Hu and Z.-C. Ou-Yang. Shape equations of the axisymmetric vesicles. *Physical Review E*, 47(1):461, 1993.

[16] F. Jülicher. The morphology of vesicles of higher topological genus: conformal degeneracy and conformal modes. *Journal de Physique II*, 6(12):1797–1824, 1996.

[17] F. Jülicher, U. Seifert, and R. Lipowsky. Conformal degeneracy and conformal diffusion of vesicles. *Physical review letters*, 71(3):452, 1993.

[18] S. Krivoshapko and V. N. Ivanov. *Encyclopedia of analytical surfaces*. Springer, New York, 2015.

[19] D. F. Lawden. *Elliptic functions and applications*. Springer, New York, 1989.

[20] K. R. Meyer. Jacobi elliptic functions from a dynamical systems point of view. *The American Mathematical Monthly*, 108(8):729–737, 2001.

[21] H. Naito, M. Okuda, and O.-Y. Zhong-Can. New solutions to the Helfrich variation problem for the shapes of lipid bilayer vesicles: Beyond Delaunay's surfaces. *Physical review letters*, 74(21):4345, 1995.

[22] P. J. Olver. *Applications of Lie groups to differential equations*, volume 107 of *Graduate Texts in Mathematics*. Springer, 2003.

[23] Z.-C. Ou-Yang and W. Helfrich. Bending energy of vesicle membranes: General expressions for the first, second, and third variation of the shape energy and applications to spheres and cylinders. *Physical Review A*, 39:5280, 1989.

[24] Z.-C. Ou-Yang, J.-X. Liu, and Y.-Z. Xie. *Geometric Methods in the Elastic Theory of Membranes in Liquid Crystal Phases*. World Scientific, Hong Kong, 1999.

[25] S. Poisson. Mémoire sur les surfaces élastiques. *Mem. Cl. Sci. Math. Phys., Inst. de France*, pages 167–225., 1812.

[26] T. Riviere. Analysis aspects of Willmore surfaces. *Inventiones mathematicae*, 174(1):1–45, 2008.

[27] W. Schief, M. Kléman, and C. Rogers. On a nonlinear elastic shell system in liquid crystal theory: generalized Willmore surfaces and Dupin cyclides. *Proceedings of the Royal Society A: Mathematical, Physical and Engineering Sciences*, 461(2061):2817–2837, Jul 2005.

[28] U. Seifert. Conformal transformations of vesicle shapes. *Journal of Physics A: Mathematical and General*, 24(11):L573, 1991.

[29] J. Stillwell. *Mathematics and Its History*. Springer, New York, 1989.

[30] G. Thomsen. Grundlagen der konformen flächentheorie. In *Abhandlungen aus dem Mathematischen Seminar der Universität Hamburg*, volume 3, pages 31–56, 1924.

[31] M. Toda and B. Athukoralage. Geometry of biological membranes and Willmore energy. In *11-th International Conference of Numerical Analysis and Applied Mathematics (ICNAAM 2013)*, volume 1558 of *AIP Conf. Proc.*, pages 883–886. AIP Publishing, 2013.

[32] M. Toda and B. Athukorallage. Geometric models for secondary structures in proteins. *Geom. Integrability & Quantization*, 16:282–300, 2015.

[33] V. M. Vassilev, P. A. Djondjorov, and I. M. Mladenov. Cylindrical equilibrium shapes of fluid membranes. *Journal of Physics A: Mathematical and Theoretical*, 41:435201, 2008.

[34] V. M. Vassilev, P. A. Djondjorov, and I. M. Mladenov. Comment on "Shape transition of unstrained flattest single-walled carbon nanotubes under pressure" [J. Appl. Phys. 115, 044512 (2014)]. *Journal of Applied Physics*, 117:196101, May 2015.

[35] V. M. Vassilev and I. M. Mladenov. Geometric symmetry groups, conservation laws and group-invariant solutions of the Willmore equation. In *Geometry, Integrability and Quantization*, volume 5, pages 246–265, 2004.

[36] J. White. A global invariant of conformal mappings in space. *Proc. Amer. Math. Soc.*, 38:162–164, 1973.

[37] E. Whittaker and G. Watson. *A Course of Modern Analysis*. Cambridge University Press, Cambridge, 1927.

[38] T. Willmore. Note on embedded surfaces. *An. Ştiinţ. Univ. "Al. I. Cuza" Iaşi Se çt. Ia Mat.*, 11:493–496, 1965.

[39] T. Willmore. *Riemannian Geometry*. Oxford University Press, Oxford, 1993.

Chapter 4

Construction of Willmore Two-Spheres Via Harmonic Maps Into $SO^+(1, n+3)/(SO^+(1,1) \times SO(n+2))$

Peng Wang

CONTENTS

Abstract

This paper aims to provide a description of totally isotropic Willmore two-spheres and their adjoint transforms. We first recall the isotropic harmonic maps which are introduced by Hélein, Xia-Shen and Ma for the study of Willmore surfaces. Then we derive a description of the normalized potential (some Lie algebra valued meromorphic 1-forms) of totally isotropic Willmore two-spheres in terms of the isotropic harmonic maps. In particular, the corresponding isotropic harmonic maps are of finite uniton type. The proof also contains a concrete way to construct examples of totally isotropic Willmore two-spheres and their adjoint transforms. As illustrations, two kinds of examples are obtained this way.

Keywords: Willmore surfaces; Isotropic Willmore two-spheres; DPW method; adjoint transform; isotropic harmonic maps.

MSC(2010): 53A30; 58E20; 53C43; 53C35

4.1 Introduction

Totally isotropic surfaces were first introduced by Calabi [7] in the study of the global geometry of minimal two-spheres. In the study of Willmore two-spheres, totally isotropic surfaces also play an important role [11, 19, 20]. Recently, Dorfmeister and Wang used the DPW method for the conformal Gauss map to study Willmore surfaces [9]. They obtained the first new Willmore two-sphere in S^6, which admits no dual surface. Along this way, Wang [25] gives a description of all totally isotropic Willmore two-spheres in S^6, in terms of the normalized potentials of their conformal Gauss maps.

Although Willmore two-spheres may have no dual surfaces, they do admit another kind of transforms, i.e., adjoint transforms introduced by Ma [15]. The main idea of the adjoint transforms is to find out another Willmore surface located in the mean curvature spheres of the original one and having the same complex coordinates. A somewhat surprising result derived by Ma states that a Willmore surface and its adjoint surface in S^{n+2} also provide a harmonic map into the Grassmannian $Gr_{1,1}(\mathbb{R}^{n+4}_1)$, which was first discovered by Hélein [13, 14] and generalized by Xia and Shen [27]. This kind of harmonic maps also appeared naturally when Brander and Wang considered the Björling problems of Willmore surfaces [1].

However, such harmonic maps will have singularities in general (See Section 6 for example). So it is very hard to use them to discuss the global geometry of Willmore surfaces. Due to this reason, in [9] Dorfmeister and Wang mainly dealt with the conformal Gauss maps of Willmore surfaces, which are another kind of harmonic maps related to Willmore surfaces globally.

Although it exists locally in general, the harmonic map given by a Willmore surface and its adjoint surface is very simple and provides the Willmore surface and its adjoint surface immediately. So it is natural to use this harmonic map to describe totally isotropic Willmore two-spheres. In particular, this provides a more simple way to derive examples of totally isotropic Willmore two-spheres and their adjoint surfaces, in contrast to using the conformal Gauss maps of Willmore surfaces [25].

In this paper, we will give a characterization of the harmonic map which is expressed by a totally isotropic Willmore two-sphere and its adjoint surface. In particular, such harmonic maps are very simple so that one can obtain a concrete algorithm to construct all of them, which is the main topic of this paper. As illustrations, we also derive two kinds of examples.

The main idea of our work is based on the DPW method for harmonic maps [8, 13] and the description of harmonic maps of finite uniton type [4, 12, 10]. The DPW method [8] gives a way to produce harmonic maps in terms of some meromorphic 1-forms, i.e., normalized potentials. The work of [4, 12, 10] states that harmonic maps of finite uniton type can be derived in a more convenient way, that is, the normalized potentials must take values in some nilpotent Lie algebra. This permits a way to derive such harmonic maps in an explicit way. On the other hand, due to [22, 4, 12, 10], harmonic maps from two-spheres into an inner symmetric space will always be of finite uniton type. This provides a way to classify all Willmore two-spheres in terms of their conformal Gauss maps [24]. For the harmonic maps used in this paper, a main problem is that they are not globally well-defined in general. So one can not apply the theory to such harmonic maps. But using Wu's formula and the description of the normalized potentials of harmonic maps of finite uniton type, we are able to show that the harmonic map given by a totally isotropic Willmore two-sphere and its adjoint surface is also of finite uniton type. So far we do not have a clear explain for this phenomenon, which may need a detailed discussion on the Iwasawa cells of the corresponding non-compact Loop groups. We hope to continue this study in a future publication.

This paper is organized as follows: In Section 2, we first recall some basic results about Willmore surfaces and their adjoint transforms. Then in Section 3 we discuss the isotropic harmonic maps given by Willmore surfaces and their adjoint transforms. Section 4 provides a description of the normalized potentials of totally isotropic Willmore two-spheres in terms of the isotropic harmonic maps. The converse part, i.e., that generically such normalized potentials will always produce totally isotropic Willmore surfaces and their adjoint transforms, is the main content of Section 5. Using these results, we also derive some concrete examples in Section 5. Then we end the paper by Section 6, which contain the technical and tedious computations of Section 5.

4.2 Willmore surfaces and adjoint surfaces

In this section we will first recall the basic surface theory of Willmore surfaces in S^{n+2} in the spirit of the treatment of [5, 16]. Then we will collect the descriptions of the adjoint transforms of Willmore surfaces [15]. We refer to [16, 15, 1] for more details.

4.2.1 Review of Willmore surfaces in S^{n+2}

Let $\mathbb{R}^{n+4}_1$ be the Minkowski space, with a Lorentzian metric $\langle x, y\rangle = -x_1y_1 + \sum_{j=2}^{n+3} x_jy_j = x^t I_{1,n+3} y$, $I_{1,n+3} = \mathrm{diag}(-1, 1, \cdots, 1)$. Let $\mathcal{C}^{n+3}_+ := \{x \in \mathbb{R}^{n+4}_1 | \langle x, x\rangle = 0, x_1 > 0\}$ be the forward light cone. Let $Q^{n+2} := \{ [x] \in \mathbb{R}P^{n+3} \mid x \in \mathcal{C}^{n+3}_+\}$ be the the projective light cone with the induced conformal metric. Then Q^{n+2} is conformally equivalent to S^{n+2}, and the conformal group of $S^{n+2} \cong Q^{n+2}$ is the orthogonal group $O(1, n+3)/\{\pm 1\}$ of $\mathbb{R}^{n+4}_1$, acting on Q^{n+2} by $T([x]) = [Tx]$ for any $T \in O(1, n+3)$. Let $SO^+(1, n+3)$ be the connected component of $O(1, n+3)$ containing I, i.e.,

$$SO^+(1, n+3) = \{T \in O(1, n+3) |\ \det T = 1,\ T \text{ preserves the time direction of } \mathbb{R}^{n+4}_1\}.$$

Let $y : M \to S^{n+2}$ be a conformal immersion from a Riemann surface M, with z a local complex coordinate on $U \subset M$ and $\langle y_z, y_{\bar z}\rangle = \frac{1}{2}e^{2\omega}$. The lift $Y : U \to \mathcal{C}^{n+3}_+$ is called a canonical lift of y with respect to z, satisfying $|\mathrm{d}Y|^2 = |\mathrm{d}z|^2$. Then there is a bundle decomposition

$$M \times \mathbb{R}^{n+4}_1 = V \oplus V^{\perp}, \text{ with } V = \mathrm{Span}_{\mathbb{R}}\{Y, \mathrm{Re}Y_z, \mathrm{Im}Y_z, Y_{z\bar z}\},\ V^{\perp} \perp V.$$

Here V is a Lorentzian rank-4 sub-bundle. This decomposition is independent of the choice of Y and z. We denote by $V_{\mathbb{C}}$ and $V_{\mathbb{C}}^{\perp}$ as their complexifications. There exists a unique section $N \in \Gamma(V)$ such that $\langle N, Y_z\rangle = \langle N, Y_{\bar{z}}\rangle = \langle N, N\rangle = 0, \langle N, Y\rangle = -1$. Noting that Y_{zz} is orthogonal to Y, Y_z and $Y_{\bar{z}}$, there exists a complex function s and a section $\kappa \in \Gamma(V_{\mathbb{C}}^{\perp})$ such that $Y_{zz} = -\frac{s}{2}Y + \kappa$. This defines two basic invariants κ and s depending on coordinates z, *the conformal Hopf differential* and *the Schwarzian* of y [5]. Let D denote the normal connection and $\psi \in \Gamma(V_{\mathbb{C}}^{\perp})$ any section of the normal bundle. The structure equations can be given as follows:

$$\begin{cases} Y_{zz} = -\dfrac{s}{2}Y + \kappa, \\ Y_{z\bar{z}} = -\langle \kappa, \bar{\kappa}\rangle Y + \dfrac{1}{2}N, \\ N_z = -2\langle \kappa, \bar{\kappa}\rangle Y_z - sY_{\bar{z}} + 2D_{\bar{z}}\kappa, \\ \psi_z = D_z\psi + 2\langle \psi, D_{\bar{z}}\kappa\rangle Y - 2\langle \psi, \kappa\rangle Y_{\bar{z}}. \end{cases}$$

The conformal Gauss, Codazzi and Ricci equations as integrable conditions are:

$$\begin{aligned} &\frac{s_{\bar{z}}}{2} = 3\langle \kappa, D_z\bar{\kappa}\rangle + \langle D_z\kappa, \bar{\kappa}\rangle, \quad \mathrm{Im}(D_{\bar{z}}D_{\bar{z}}\kappa + \frac{\bar{s}}{2}\kappa) = 0, \\ &D_{\bar{z}}D_z\psi - D_zD_{\bar{z}}\psi = 2\langle \psi, \kappa\rangle\bar{\kappa} - 2\langle \psi, \bar{\kappa}\rangle\kappa. \end{aligned} \tag{4.1}$$

The conformal Gauss map of y is defined as follows.

Definition 4.1. *[3, 5, 11, 16] For a conformally immersed surface* $y : M \to S^{n+2}$, the conformal Gauss map $Gr(p) : M \to Gr_{3,1}(\mathbb{R}^{n+4}_1) = SO^+(1, n+3)/(SO^+(1,3) \times SO(n))$ *of* y *is defined as*

$$Gr(p) := V_p.$$

So locally we have $Gr = Y \wedge Y_u \wedge Y_v \wedge N = -2i \cdot Y \wedge Y_z \wedge Y_{\bar{z}} \wedge N$, *with* $z = u + iv$.

Direct computation shows that Gr induces a conformal-invariant metric $\mathbf{g} := \frac{1}{4}\langle \mathrm{d}Gr, \mathrm{d}Gr\rangle = \langle \kappa, \bar{\kappa}\rangle|\mathrm{d}z|^2$ on M. Note $\mathbf{g}$ degenerates at umibilic points of y. The Willmore functional and Willmore surfaces can be defined by use of this metric.

Definition 4.2. The Willmore functional *of* y *is defined as:*

$$W(y) := 2i \int_M \langle \kappa, \bar{\kappa}\rangle \mathrm{d}z \wedge \mathrm{d}\bar{z}.$$

An immersed surface $y : M \to S^{n+2}$ *is called a* Willmore surface *if it is a critical surface of the Willmore functional with respect to any variation of the map* $y : M \to S^{n+2}$.

It is well-known that [3, 5, 11, 23] y is Willmore if and only if

$$D_{\bar{z}}D_{\bar{z}}\kappa + \frac{\bar{s}}{2}\kappa = 0;$$

if and only if the conformal Gauss map $Gr : M \to Gr_{3,1}(\mathbb{R}^{n+3}_1)$ is harmonic. We refer to [9] for the conformal Gauss map approach for Willmore surface.

4.2.2 Adjoint transforms of a Willmore surface and the second harmonic map related to Willmore surfaces

Transforms play an important role in the study of Willmore surfaces. For a Willmore surface y in S^3, it was shown by Bryant in the seminal paper [3] that they always admit a unique dual surface which may have branch points or degenerate to a point. Hence the dual surface is either degenerate or has the same complex coordinate and the same conformal Gauss map as y at the points it is immersed. This duality theorem, however, does not hold in general when the codimension is bigger than 1 ([11], [5], [15]). To characterize Willmore surfaces with dual surfaces, in [11] Ejiri introduced the notion of *S-Willmore surfaces*. Here we define it slightly differently to include all Willmore surfaces with dual surfaces:

Definition 4.3. *A Willmore immersion $y : M^2 \to S^{n+2}$ is called an S-Willlmore surface if its conformal Hopf differential satisfies*

$$D_{\bar{z}}\kappa || \kappa,$$

i.e. there exists some function μ on M such that $D_{\bar{z}}\kappa + \frac{\mu}{2}\kappa = 0$.

A basic result of [11] states that a Willmore surface admits a dual surface if and only if it is S-Willmore. Moreover the dual surface is also Willmore at the points it is immersed.

To consider the generic Willmore surfaces, Ma introduced the adjoint transform of a Willmore surface y [15, 16]. An adjoint transform of y is a conformal map $\hat{y}$ which is located on the mean curvature sphere of y and satisfies some additional condition. To be concrete we have the following

4.2.2.1 Adjoint transforms

Let $y : U \to S^{n+2}$ be an umbilic free Willmore surface with canonical lift Y with respect to z as above. Set

$$\hat{Y} = N + \bar{\mu}Y_z + \mu Y_{\bar{z}} + \frac{1}{2}|\mu|^2 Y, \tag{4.2}$$

with $\mu \mathrm{d}z = 2\langle \hat{Y}, Y_z\rangle \mathrm{d}z$ a connection 1–form. Direct computation yields [15]

$$\hat{Y}_z = \frac{\mu}{2}\hat{Y} + \theta\left(Y_{\bar{z}} + \frac{\bar{\mu}}{2}Y\right) + \rho\left(Y_z + \frac{\mu}{2}Y\right) + 2\zeta \tag{4.3}$$

with

$$\theta := \mu_z - \frac{\mu^2}{2} - s, \ \rho := \bar{\mu}_z - 2\langle \kappa, \bar{\kappa}\rangle, \ \zeta := D_{\bar{z}}\kappa + \frac{\bar{\mu}}{2}\kappa.$$

Now we define the adjoint surface as follows:

Definition 4.4. *[15] The map $\hat{Y} : U \to S^{n+2}$ is called an adjoint transform of the Willmore surface Y if the following two equations hold for μ:*

$$\mu_z - \frac{\mu^2}{2} - s = 0, \ \textit{Riccati equation}, \tag{4.4}$$

$$\langle D_{\bar{z}}\kappa + \frac{\bar{\mu}}{2}\kappa, D_{\bar{z}}\kappa + \frac{\bar{\mu}}{2}\kappa\rangle = 0. \tag{4.5}$$

Note that $\hat{Y}$ is the dual surface of Y if and only if $D_{\bar{z}}\kappa + \frac{\bar{\mu}}{2}\kappa = 0$ ([3], [11], [15]).

Theorem 4.5. *[15]* Willmore property and existence of adjoint transform: *The adjoint transform $\hat{Y}$ of a Willmore surface y is also a Willmore surface (may degenerate). Moreover,*

1. *If $\langle \kappa, \kappa\rangle \equiv 0$, any solution to the equation (4.4) is also a solution to the equation (4.5). Hence, there exist infinitely many adjoint surfaces of y in this case.*
2. *If $\langle \kappa, \kappa\rangle \neq 0$ and $\Omega \mathrm{d}z^6 := \langle D_{\bar{z}}\kappa, \kappa\rangle^2 - \langle \kappa, \kappa\rangle\langle D_{\bar{z}}\kappa, D_{\bar{z}}\kappa\rangle \mathrm{d}z^6 \neq 0$, there are exactly two different solutions to equation (4.5), which also solve (4.4). Hence, there exist exactly two adjoint surfaces of y in this case.*
3. *If $\langle \kappa, \kappa\rangle \neq 0$ and $\langle D_{\bar{z}}\kappa, \kappa\rangle^2 - \langle \kappa, \kappa\rangle\langle D_{\bar{z}}\kappa, D_{\bar{z}}\kappa\rangle \equiv 0$, there exists a unique solution to (4.5), which also solves (4.4). Hence, there exists a unique adjoint surface of y in this case.*

Remark 4. *In [6], dressing transformations of constrained Willmore surfaces are discussed in details. It stays unclear whether the adjoint transforms can be derived as a special kind of dressing transformations.*

4.2.2.2 Harmonic maps into $SO^+(1, n+3)/(SO^+(1,1) \times SO(n+2))$ related to Willmore surfaces

A crucial observation by Hélein etc. [13, 14, 27, 15] is that Y and $\hat{Y}$ produce furthermore a second useful harmonic map related to a Willmore surface y.

Theorem 4.6. *Let $[Y]$ be a Willmore surface. Let μ be a solution to the Riccati equation (4.4) on U, defining $\hat{Y}$ as (4.2). Let $\mathcal{F}_h : U \to SO^+(1, n+3)/(SO^+(1,1) \times SO(n+2))$ be the map taking p to $Y(p) \wedge \hat{Y}(p)$. We have the following results.*

1. *([13, 14, 27]) The map $\mathcal{F}_h$ is harmonic, and is called a* half–isotropic harmonic map with respect to Y.

2. *([15]) If μ also solves (4.5), i.e., $\hat{Y}$ is an adjoint transform of y, then $\mathcal{F}_h$ is* conformally *harmonic, and is called an* isotropic harmonic map with respect to Y.

At umbilic points it is possible that there exists a limit of μ such that (4.4) holds. Due to the following lemma, the harmonic map $\mathcal{F}_h$ has no definition when μ tends to ∞.

Lemma 4.7. *[11, 9] At the umbilic points of Y, the limit of μ goes to a finite number or infinity. When μ goes to infinity, $[\hat{Y}]$ tends to $[Y]$, and at the limit point we have $[\hat{Y}] = [Y]$.*

Restricting to the isotropic harmonic map, we have the following description.

Theorem 4.8. *[15], [14], [1] Let $\mathcal{F}_h = Y \wedge \hat{Y}$ be an isotropic harmonic map. Set $e_1, e_2 \in \Gamma(V)$ with $Y_z + \frac{\mu}{2}Y = \frac{1}{2}(e_1 - ie_2)$. Let $\{\psi_j, j = 1, \cdots, n\}$ be a frame of the normal bundle $V^{\perp}$. Assume that $\kappa = \sum_{j=1}^n k_j\psi_j$, $\zeta = \sum_{j=1}^n \gamma_j\psi_j$, $D_z\psi_j = \sum_{l=1}^n b_{jl}\psi_l$, $b_{jl} + b_{lj} = 0$. Set*

$$F = \left(\frac{1}{\sqrt{2}}(Y + \hat{Y}), \frac{1}{\sqrt{2}}(-Y + \hat{Y}), e_1, e_2, \psi_1, \cdots, \psi_n\right).$$

Then the Maurer-Cartan form $\alpha = F^{-1}\mathrm{d}F = \alpha' + \alpha''$ of F has the structure:

$$\alpha' = \begin{pmatrix} A_1 & B_1 \\ -B_1^t I_{1,1} & A_2 \end{pmatrix} \mathrm{d}z,$$

with

$$A_1 = \begin{pmatrix} 0 & \frac{\mu}{2} \\ \frac{\mu}{2} & 0 \end{pmatrix}, \ B_1 = \begin{pmatrix} \frac{1+\rho}{2\sqrt{2}} & \frac{-i-i\rho}{2\sqrt{2}} & \sqrt{2}\gamma_1 & \cdots & \sqrt{2}\gamma_n \\ \frac{1-\rho}{2\sqrt{2}} & \frac{-i+i\rho}{2\sqrt{2}} & -\sqrt{2}\gamma_1 & \cdots & -\sqrt{2}\gamma_n \end{pmatrix} = \begin{pmatrix} b_1^t \\ b_2^t \end{pmatrix},$$

and

$$B_1 B_1^t = 0. \tag{4.6}$$

Conversely, if $\mathcal{F} = Y \wedge \hat{Y} : U \to SO^+(1, n+3)/(SO^+(1,1) \times SO(n+2))$ is a conformal harmonic map satisfying (4.6), then $\mathcal{F}$ is an isotropic harmonic map and Y and $\hat{Y}$ form a pair of adjoint Willmore surfaces at the points they are immersed. Moreover, set

$$B_1 = (b_1 \ b_2)^t \ \text{with} \ b_1, b_2 \in \mathbb{C}^{n+2}.$$

Then Y is immersed at the points $(b_1^t + b_2^t)(\bar{b}_1 + \bar{b}_2) > 0$ and $\hat{Y}$ is immersed at the points $(b_1^t - b_2^t)(\bar{b}_1 - \bar{b}_2) > 0$.

4.3 Isotropic harmonic maps into $SO^+(1,n+3)/(SO^+(1,1)\times SO(n+2))$

In this section we will recall briefly the DPW construction of harmonic maps and applications to the isotropic harmonic maps related to Willmore surfaces. We refer to [13, 14, 27, 1] for more details.

4.3.1 The DPW construction of harmonic maps

4.3.1.1 Harmonic maps into an inner symmetric space

Let G/K be an inner symmetric space with involution $\sigma : G \to G$ such that $G^\sigma \supset K \supset (G^\sigma)_0$. Let $\pi : G \to G/K$ be the projection of G into G/K. Let $\mathfrak{g} = Lie(G)$ and $\mathfrak{k} = Lie(K)$ be their Lie algebras. We have the Cartan decomposition $\mathfrak{g} = \mathfrak{k} \oplus \mathfrak{p}$, $[\mathfrak{k},\mathfrak{k}] \subset \mathfrak{k}$, $[\mathfrak{k},\mathfrak{p}] \subset \mathfrak{p}$, $[\mathfrak{p},\mathfrak{p}] \subset \mathfrak{k}$.

Let $\mathcal{F} : M \to G/K$ be a conformal harmonic map from a Riemann surface M, with $U \subset M$ an open, simply connected subset. Then there exists a frame $F : U \to G$ such that $\mathcal{F} = \pi \circ F$. So we have the Maurer-Cartan form $F^{-1}\mathrm{d}F = \alpha$, and Maurer-Cartan equation $\mathrm{d}\alpha + \frac{1}{2}[\alpha \wedge \alpha] = 0$. Set $\alpha = \alpha_0 + \alpha_1$, with $\alpha_0 \in \Gamma(\mathfrak{k} \otimes T^*M)$, $\alpha_1 \in \Gamma(\mathfrak{p} \otimes T^*M)$. Decompose α_1 further into the $(1,0)-$part α_1' and the $(0,1)-$part α_1''. Then set $\alpha_\lambda = \lambda^{-1}\alpha_1' + \alpha_0 + \lambda\alpha_1''$, with $\lambda \in S^1$. We have the well-known characterization of harmonic maps:

Lemma 4.9. *([8]) The map $\mathcal{F} : M \to G/K$ is harmonic if and only if*

$$\mathrm{d}\alpha_\lambda + \frac{1}{2}[\alpha_\lambda \wedge \alpha_\lambda] = 0 \quad \textit{for all } \lambda \in S^1.$$

As a consequence, for a harmonic map f, the equation $\mathrm{d}F(z,\lambda) = F(z,\lambda)\,\alpha_\lambda$ with $F(0,\lambda) = F(0)$, always has a solution, which is called the *extended frame* of $\mathcal{F}$.

4.3.1.2 Two decomposition theorems

We denote by $SO^+(1,n+3)$ the connected component of the identity of the linear isometry group of $\mathbb{R}^{n+4}_1$. Then

$$\mathfrak{so}(1,n+3) = \mathfrak{g} = \{X \in \mathfrak{gl}(n+4,\mathbb{R}) | X^t I_{1,n+3} + I_{1,n+3}X = 0\}.$$

Define the involution

$$\begin{array}{ll} \sigma : SO^+(1,n+3) & \to SO^+(1,n+3) \\ \qquad A & \mapsto DAD^{-1}, \end{array} \qquad \text{where} \quad D = \begin{pmatrix} -I_2 & 0 \\ 0 & I_{n+2} \end{pmatrix}.$$

We have $SO^+(1,n+3)^\sigma \supset SO^+(1,1) \times SO(n+2) = (SO^+(1,n+3)^\sigma)_0$. We also have

$$\mathfrak{g} = \left\{ \begin{pmatrix} A_1 & B_1 \\ -B_1^t I_{1,1} & A_2 \end{pmatrix} | A_1^t I_{1,1} + I_{1,1}A_1 = 0, A_2 + A_2^t = 0 \right\} = \mathfrak{k} \oplus \mathfrak{p},$$

with

$$\mathfrak{k} = \left\{ \begin{pmatrix} A_1 & 0 \\ 0 & A_2 \end{pmatrix} | A_1^t I_{1,1} + I_{1,1} A_1 = 0, A_2 + A_2^t = 0 \right\},$$
$$\mathfrak{p} = \left\{ \begin{pmatrix} 0 & B_1 \\ -B_1^t I_{1,1} & 0 \end{pmatrix} \right\}.$$

Let

$$G^{\mathbb{C}} = SO^+(1, n+3, \mathbb{C}) := \{X \in SL(n+4, \mathbb{C}) \mid X^t I_{1,n+3} X = I_{1,n+3}\},$$
$$\mathfrak{g}^{\mathbb{C}} = \mathfrak{so}(1, n+3, \mathbb{C}).$$

Extend σ to an inner involution of $G^{\mathbb{C}}$ with fixed point group $K^{\mathbb{C}} = S(O^+(1,1,\mathbb{C}) \times O(n+2, \mathbb{C}))$.

Let $\Lambda G_\sigma^{\mathbb{C}}$ denote the group of loops in $G^C = SO^+(1, n+3, \mathbb{C})$ twisted by σ. Let $\Lambda^+ G_\sigma^{\mathbb{C}}$ denote the subgroup of loops which extend holomorphically to the unit disk $|\lambda| \leq 1$. We also need the subgroup $\Lambda_B^+ G_\sigma^{\mathbb{C}} := \{\gamma \in \Lambda^+ G_\sigma^{\mathbb{C}} \mid \gamma|_{\lambda=0} \in \mathfrak{B}\}$, where $\mathfrak{B} \subset K^{\mathbb{C}}$ is defined from the Iwasawa decomposition $K^{\mathbb{C}} = K \cdot \mathfrak{B}$. In this case,

$$\mathfrak{B} = \left\{ \begin{pmatrix} \mathrm{b}_1 & 0 \\ 0 & \mathrm{b}_2 \end{pmatrix} \middle| \mathrm{b}_1 = \begin{pmatrix} \cos\theta & i\sin\theta \\ i\sin\theta & \cos\theta \end{pmatrix}, \right.$$
$$\left. \theta \in \mathbb{R} \mod 2\pi\mathbb{Z}, \text{ and } \mathrm{b}_2 \in \mathfrak{B}_2 \right\}.$$

Here $\mathfrak{B}_2$ is the solvable subgroup of $SO(n+2, \mathbb{C})$ such that $SO(n+2, \mathbb{C}) = SO(n+2) \cdot \mathfrak{B}_2$. We refer to Lemma 4 of [13] for more details.

Theorem 4.10. *Theorem 5 of [13], see also [27], [8], [21] (Iwasawa decomposition): The multiplication $\Lambda G_\sigma \times \Lambda_B^+ G^{\mathbb{C}} \to \Lambda G_\sigma^{\mathbb{C}}$ is a real analytic diffeomorphism onto the open dense subset $\Lambda G_\sigma \cdot \Lambda_B^+ G^{\mathbb{C}} \subset \Lambda G_\sigma^{\mathbb{C}}$.*

Let $\Lambda_*^- G_\sigma^{\mathbb{C}}$ denote the loops that extend holomorphically into ∞ and take values I at infinity.

Theorem 4.11. *Theorem 7 of [13], see also [27], [8], [21] (Birkhoff decomposition): The multiplication $\Lambda_*^- G_\sigma^{\mathbb{C}} \times \Lambda^+ G^{\mathbb{C}} \to \Lambda G_\sigma^{\mathbb{C}}$ is a real analytic diffeomorphism onto the open subset $\Lambda_*^- G_\sigma^{\mathbb{C}} \cdot \Lambda^+ G^{\mathbb{C}}$ (the big cell) of $\Lambda G_\sigma^{\mathbb{C}}$.*

4.3.1.3 The DPW construction and Wu's formula

Here we recall the DPW construction for harmonic maps. Let $\mathbb{D} \subset \mathbb{C}$ be a disk or $\mathbb{C}$ itself, with complex coordinate z.

Theorem 4.12. *[8]*

1. *Let $\mathcal{F} : \mathbb{D} \to G/K$ be a harmonic map with an extended frame $F(z,\bar{z},\lambda) \in \Lambda G_\sigma$ and $F(0,0,\lambda) = I$. Then there exists a Birkhoff decomposition*
$$F_-(z,\lambda) = F(z,\bar{z},\lambda)F_+(z,\bar{z},\lambda), \quad \text{with} \;\; F_+ \in \Lambda^+ G_\sigma^{\mathbb{C}},$$
such that $F_-(z,\lambda) : \mathbb{D} \to \Lambda_*^- G_\sigma^{\mathbb{C}}$ is meromorphic. Moreover, the Maurer-Cartan form of F_- is of the form
$$\eta = F_-^{-1}\mathrm{d}F_- = \lambda^{-1}\eta_{-1}(z)\mathrm{d}z,$$
with η_{-1} independent of λ. The 1-form η is called the normalized potential of $\mathcal{F}$.

2. *Let η be a $\lambda^{-1} \cdot \mathfrak{p}-$valued meromorphic 1-form on $\mathbb{D}$. Let $F_-(z,\lambda)$ be a solution to $F_-^{-1}\mathrm{d}F_- = \eta$, $F_-(0,\lambda) = I$. Then on an open subset $\mathbb{D}_\mathfrak{I}$ of $\mathbb{D}$ one has*
$$F_-(z,\lambda) = \tilde{F}(z,\bar{z},\lambda) \cdot \tilde{F}_+(z,\bar{z},\lambda), \quad \text{with} \;\; \tilde{F} \in \Lambda G_\sigma, \;\; \tilde{F}_+ \in \Lambda_B^+ G_\sigma^{\mathbb{C}}.$$
This way, one obtains an extended frame $\tilde{F}(z,\bar{z},\lambda)$ of some harmonic map from $\mathbb{D}_\mathfrak{I}$ to G/K with $\tilde{F}(0,\lambda) = I$. Moreover, all harmonic maps can be obtained in this way, since these two procedures are inverse to each other if the normalization at some based point is used.

The normalized potential can be determined in the following way. Let f and F be as above. Let $\alpha_\lambda = F^{-1}\mathrm{d}F$. Let δ_1 and δ_0 denote the sum of the holomorphic terms of z about $z = 0$ in the Taylor expansion of $\alpha_1'(\frac{\partial}{\partial z})$ and $\alpha_0'(\frac{\partial}{\partial z})$.

Theorem 4.13. *[26] (Wu's formula) We retain the notations in Theorem 4.12. Then the normalized potential of $\mathcal{F}$ with respect to the base point 0 is given by*
$$\eta = \lambda^{-1}F_0(z)\delta_1 F_0(z)^{-1}\mathrm{d}z, \tag{4.7}$$
where $F_0(z) : \mathbb{D} \to G^{\mathbb{C}}$ is the solution to $F_0(z)^{-1}\mathrm{d}F_0(z) = \delta_0\mathrm{d}z$, $F_0(0) = I$.

Lemma 4.14. *We retain the notations in Theorem 4.13. Let $Q \in K$ and $Q\mathcal{F}$ be a transform of $\mathcal{F}$ in G/K. Then the normalized potential of $Q\mathcal{F}$ with respect to the base point 0 is*
$$\eta_Q = Q\eta Q^{-1}.$$

Proof. We have now the lift QFQ^{-1} of $Q\mathcal{F}$ with respect to the base point 0. So we have the Birkhoff splitting of QFQ^{-1} as below
$$F_{-Q} = QF_-Q^{-1} = QFQ^{-1}QF_+Q^{-1} \text{ since } F_- = FF_+.$$
Hence we obtain
$$\eta_Q = (QF_-Q^{-1})^{-1}\mathrm{d}QF_-Q^{-1} = Q\eta Q^{-1}.$$
□

This lemma shows that we can identify the normalized potentials up to an conjugation of elements in K.

4.3.2 Potentials of isotropic harmonic maps

4.3.2.1 The general case

Let $\mathbb{D}$ denote the unit disk of $\mathbb{C}$ or $\mathbb{C}$ itself. Let $\mathcal{F}:\mathbb{D}\rightarrow SO^+(1,n+3)/(SO^+(1,1)\times SO(n+2))$ be a harmonic map with a lift $F:\mathbb{D}\rightarrow SO^+(1,n+3)$ and the Maurer-Cartan form $\alpha=F^{-1}\mathrm{d}F$. Then

$$\alpha_0'=\left(\begin{array}{cc} A_1 & 0\\ 0 & A_2\end{array}\right)\mathrm{d}z,\ \ \alpha_1'=\left(\begin{array}{cc} 0 & B_1\\ -B_1^tI_{1,1} & 0\end{array}\right)\mathrm{d}z.$$

Theorem 4.15. *([13, 14, 27, 1]) The normalized potential of an isotropic harmonic map $\mathcal{F}=Y\wedge\hat{Y}$ is of the form*

$$\eta=\lambda^{-1}\left(\begin{array}{cc} 0 & \hat{B}_1\\ -\hat{B}_1^tI_{1,1} & 0\end{array}\right)\mathrm{d}z,\ \ with\ \ \hat{B}_1\hat{B}_1^t=0. \tag{4.8}$$

Moreover, $[Y]$ and $[\hat{Y}]$ forms a pair of dual (S-)Willmore surfaces if and only if $rank(\hat{B}_1)=1$.

Conversely, let $\mathcal{F}=Y\wedge\hat{Y}$ be an harmonic map with normalized potential

$$\eta=\lambda^{-1}\left(\begin{array}{cc} 0 & \hat{B}_1\\ -\hat{B}_1^tI_{1,1} & 0\end{array}\right)\mathrm{d}z$$

satisfying (4.8). Then $\mathcal{F}$ is an isotropic harmonic map.

4.3.2.2 On minimal surfaces in space forms

In [13], there is an interesting description of Willmore surfaces Möbius equivalent to minimal surfaces in space forms. Here we restate it as:

Theorem 4.16. *([13], [27]) Let $\mathcal{F}_h=Y\wedge\hat{Y}$ be a non-constant isotropic harmonic map.*

1. *The map $[Y]$ is Möbius equivalent to a minimal surface in $\mathbb{R}^{n+2}$ if $\hat{Y}$ reduces to a point. In this case $B_1=\left(\begin{array}{cc} b_1 & b_1\end{array}\right)^t$.*

2. *The map $[Y]$ is Möbius equivalent to a minimal surface in S^{n+2} if $\mathcal{F}_h$ reduces to a harmonic map into $SO(n+3)/SO(n+2)$. In this case $B_1=\left(\begin{array}{cc} 0 & b_1\end{array}\right)^t$.*

3. *The map $[Y]$ is Möbius equivalent to a minimal surface in H^{n+2} if $\mathcal{F}_h$ reduces to a harmonic map into $SO^+(1,n+2)/SO^+(1,n+1)$. In this case $B_1=\left(\begin{array}{cc} b_1 & 0\end{array}\right)^t$.*

Here b_1 takes values in $\mathbb{C}^{n+2}$ and satisfies $b_1^tb_1=0$.

The converse of the above results also hold. That is, if B_1 is (up to conjugation) of the form stated above, then $[Y]$ is Möbius equivalent to the corresponding minimal surface where it is an immersion.

4.3.3 On harmonic maps of finite uniton type

In this subsection we will discuss harmonic maps of finite uniton type.

Loops which have a finite Fourier expansion will be called *algebraic loops* and the corresponding spaces will be denoted by the subscript "*alg*", like $\Lambda_{alg}G_\sigma$, $\Lambda_{alg}G_\sigma^{\mathbb{C}}$, $\Omega_{alg}G_\sigma$. We define

$$\Omega^k_{alg}G_\sigma := \{\gamma \in \Omega_{alg}G_\sigma | Ad(\gamma) = \sum_{|j|\leq k} \lambda^j T_j\} \ .$$

Let G/K be an inner symmetric space (given by the inner involution $\sigma : G \to G$). We map G/K into G as totally geodesic submanifold via the (finite covering) Cartan map: $\mathfrak{C} : G/K \to G, \mathfrak{C}(gK) = g\sigma(g)^{-1}$.

Definition 4.17. ([22, 4, 9, 10])

1. *Let $f : M \to G$ be a harmonic map into a real Lie group G with extended solution $\Phi(z,\lambda) \in \Lambda G_\sigma^{\mathbb{C}}$. We say that f has finite uniton number k if*

$$\Phi(M) \subset \Omega^k_{alg}G_\sigma, \quad \textit{and } \Phi(M) \not\subseteq \Omega^{k-1}_{alg}G_\sigma.$$

2. *A harmonic map f into G/K is said to be of finite uniton number k, if it is of finite uniton number k, when considered as a harmonic map into G via the Cartan map, i.e., f has finite uniton number k if and only if $\mathfrak{C} \circ f$ has finite uniton number k.*

It is proved that for harmonic maps into inner symmetric space G/K, it is of finite uniton number if and only if its normalized potential takes value in some nilpotent Lie sub-algebra [4, 12, 10]. In Section 4 and Section 5 we will give a characterization of totally isotropic Willmore two-spheres in terms of harmonic maps of finite uniton number at most 2.

4.4 Totally isotropic Willmore two-spheres and their adjoint transforms

In this section we will first collect the geometric results concerning totally isotropic Willmore two-spheres and their adjoint transforms. Then by the geometric descriptions, we are able to derive the normalized potentials of the isotropic harmonic map given by such Willmore surfaces and their adjoint transforms.

4.4.1 Totally isotropic Willmore surfaces

Let $y : M \to S^{2m}$ be a conformal immersion and we retain the notion in Section 2. Then [7, 11] y is called *totally isotropic* if and only if all the derivatives of y with respect to z are isotropic, or equivalently,

$$\langle Y_z^{(j)}, Y_z^{(l)} \rangle = 0 \quad \text{for all } j,\ l \in \mathbb{Z}^+.$$

Here "$Y_z^{(j)}$" denotes taking j times derivatives of Y by z. As a consequence a totally isotropic Willmore surface always locates in an even dimensional sphere. Moreover, we can find locally an isotropic frame $\{E_j\}$, $j = 1, \cdots, m$, such that

$$\begin{aligned}
&\langle E_j, \bar{E}_l \rangle = 2\delta_{jl},\ j, l = 1, \cdots, m, \\
&Y_z \in Span_{\mathbb{C}}\{E_1\} \mod Y, \\
&Y_z^{(j)} \in Span_{\mathbb{C}}\{E_j\} \mod Span_{\mathbb{C}}\{Y, E_1, \cdots, E_{j-1}\},\ j = 2, \cdots, m, \\
&\{E_j, \bar{E}_j\}_{j=2,\cdots,m} \text{ forms a basis of the normal bundle } V^{\perp}.
\end{aligned} \tag{4.9}$$

Next we call y an *H-totally isotropic surface* if it satisfies furthermore the following conditions

$$D_{\bar{z}} E_j \in Span_{\mathbb{C}}\{E_2, \cdots, E_m\}, \text{ for all } j = 2, \cdots, m. \tag{4.10}$$

It is direct to verify that this condition is independent of the choice of z, Y and E_j. Here the notion "H" comes from two facts. First, this condition is similar to the horizontal conditions for minimal two-spheres in S^{2m} [2, 7]. Second, by a result of [16], we can prove that

Theorem 4.18. *[18] Let y be a totaly isotropic Willmore two-sphere in S^{2m}. Then y is an H-totally isotropic surface.*

See [25] for a proof in the case $m = 3$. The key point of the proof is an application of the holomorphic forms given by κ and its derivatives, which can be found in Section 5 of [16].

Moreover, totally isotropic surfaces in S^{2m} may not be Willmore. But H-totally isotropic surfaces must be Willmore.

Proposition 4.19. *Let y be an H-totally isotropic surface in S^{2m}. Then y is Willmore.*

Proof. By definition of E_j, we have $\kappa \in Span_{\mathbb{C}}\{E_2\}$. From (4.10), we have that

$$D_{\bar{z}}\kappa \in Span_{\mathbb{C}}\{E_2, \cdots, E_m\},\ D_{\bar{z}}D_{\bar{z}}\kappa \in Span_{\mathbb{C}}\{E_2, \cdots, E_m\}.$$

So $D_{\bar{z}}D_{\bar{z}}\kappa + \frac{\bar{s}}{2}\kappa \in Span_{\mathbb{C}}\{E_2, \cdots, E_m\}$. Hence $Im(D_{\bar{z}}D_{\bar{z}}\kappa + \frac{\bar{s}}{2}\kappa) = 0$ in (4.1) indicates that the Willmore equation $D_{\bar{z}}D_{\bar{z}}\kappa + \frac{\bar{s}}{2}\kappa = 0$ holds, i.e., y is Willmore. ∎

Concerning the adjoint surfaces of H-totally isotropic surfaces, we have

Proposition 4.20. *[18] Let y be an H-totally isotropic surface in S^{2m} (hence Willmore). Then the adjoint surface of y is also H-totally isotropic surface on the points it is immersed.*

This can be easily derived since $\hat{Y}_z \in Span_{\mathbb{C}}\{E_1, \cdots, E_m\} \mod \{Y, \hat{Y}\}$ by (4.3).

4.4.2 Normalized potentials of H-totally isotropic surfaces

The normalized potentials of H-totally isotropic surfaces can be derived from Wu's formula as below

Theorem 4.21. *Let $y : \mathbb{D} \to S^{2m}$ be an H-totally isotropic surface with a local adjoint transform $\hat{y} = [\hat{Y}]$. Assume that $\mathcal{F}|_{z=0} = I \mod K$. Then up to an conjugation, the normalized potential of $\mathcal{F} = Y \wedge \hat{Y}$ has the form*

$$\eta = \lambda^{-1} \begin{pmatrix} 0 & \hat{B}_1 \\ -\hat{B}_1^t I_{1,1} & 0 \end{pmatrix} \mathrm{d}z, \quad \hat{B}_1 = \begin{pmatrix} h_{11} & ih_{11} & \cdots & h_{m1} & ih_{m1} \\ \hat{h}_{11} & i\hat{h}_{11} & \cdots & \hat{h}_{m1} & i\hat{h}_{m1} \end{pmatrix}, \tag{4.11}$$

with $\{h_{j1}\mathrm{d}z,\ \hat{h}_{j1}\mathrm{d}z |\ j = 1, \cdots, m\}$ being meromorphic 1-forms on $\mathbb{D}$.

Proof. We have the following

$$Y_z = -\frac{\mu}{2}Y + \frac{1}{2}E_1.$$

We consider the lift $\tilde{F}$ as below

$$\tilde{F} = \left(\frac{1}{\sqrt{2}}(Y + \hat{Y}), \frac{1}{\sqrt{2}}(-Y + \hat{Y}), e_1, \psi_2, \cdots, \psi_m, \hat{e}_1, \hat{\psi}_2, \cdots, \hat{\psi}_m\right).$$

Here we use the frame defined in (4.9) and set

$$E_1 = e_1 + i\hat{e}_1,\ E_j = \psi_j + i\hat{\psi}_j,\ j = 2, \cdots, m.$$

Set

$$\kappa = \sum_j k_j(\psi_j + i\hat{\psi}_j),\ \zeta = \sum_j \gamma_j(\psi_j + i\hat{\psi}_j).$$

Assume that $D_z E_j = \sum a_{jl} E_l$, $D_z \bar{E}_j = \sum \hat{a}_{jl} \bar{E}_l$. By (4.9), we have $a_{jl} + \hat{a}_{lj} = 0$, and

$$D_z \psi_j = \frac{1}{2}\left(\sum (a_{jl} - a_{lj})\psi_l + \sum i(a_{jl} + a_{jl})\hat{\psi}_l\right),$$

$$D_z \hat{\psi}_j = \frac{1}{2}\left(-\sum i(a_{jl} + a_{lj})\psi_l + \sum (a_{jl} - a_{lj})\hat{\psi}_l\right).$$

So

$$A_2 = \begin{pmatrix} A_{21} & iA_{22} \\ -iA_{22}^t & A_{21} \end{pmatrix} \quad \text{with} \quad A_{12} + A_{12}^t = 0,\ A_{22} = A_{22}^t,$$

and

$$A_{21} = \begin{pmatrix} 0 & -\frac{k_l}{2} \\ \frac{k_j}{2} & \frac{a_{jl}-a_{lj}}{2} \end{pmatrix}_{m\times m}, \quad A_{22} = \begin{pmatrix} -\frac{\mu}{2} & -\frac{k_l}{2} \\ \frac{k_j}{2} & \frac{a_{jl}+a_{lj}}{2} \end{pmatrix}_{m\times m}.$$

Hence, without lose of generality, we assume that $\tilde{F}(0,0,\lambda) = I$ and let

$$\delta_0 = \begin{pmatrix} \check{A}_1 & 0 \\ 0 & \check{A}_2 \end{pmatrix}, \quad \delta_1 = \begin{pmatrix} 0 & \check{B}_1 \\ -\check{B}_1^t I_{1,1} & 0 \end{pmatrix}$$

be the holomorphic parts of $\tilde{\alpha}_0'(\frac{\partial}{\partial z})$ and $\tilde{\alpha}_1'(\frac{\partial}{\partial z})$ respectively. Then we have

$$\check{A}_2 = \begin{pmatrix} \check{A}_{21} & i\check{A}_{22} \\ -i\check{A}_{22}^t & \check{A}_{21} \end{pmatrix} \quad \text{with } \check{A}_{21}^t + \check{A}_{21} = 0, \; \check{A}_{22}^t = \check{A}_{22}, \tag{4.12}$$

and

$$\check{B}_1 = \begin{pmatrix} \mathbf{b}_1^t & i\mathbf{b}_1^t \\ \hat{\mathbf{b}}_1^t & i\hat{\mathbf{b}}_1^t \end{pmatrix}.$$

Let $F_0(z) : \mathbb{D} \to G^{\mathbb{C}}$ be the solution to $F_0(z)^{-1}\mathrm{d}F_0(z) = \delta_0 \mathrm{d}z$, $F_0(0) = I$. We see that

$$F_0 = \begin{pmatrix} F_{01} & 0 \\ 0 & F_{02} \end{pmatrix}$$

with $F_{01} = \exp(z\check{A}_1)$ and

$$F_{02} = \exp(z\check{A}_2) = \begin{pmatrix} F_{021} & F_{022} \\ -F_{022} & F_{021} \end{pmatrix}, \quad F_{02}^{-1} = F_{02}^t = \begin{pmatrix} F_{021}^t & -F_{022}^t \\ F_{022}^t & F_{021}^t \end{pmatrix},$$

since $\check{A}_2$ satisfies (4.12). So the normalized potential has the form by Wu's formula (4.7)

$$\widetilde{\eta} = \lambda^{-1} \begin{pmatrix} 0 & \widetilde{B}_1 \\ -\widetilde{B}_1^t I_{1,1} & 0 \end{pmatrix} \mathrm{d}z \quad \text{with} \quad \widetilde{B}_1 = F_{01}\check{B}_1 F_{02}^{-1}.$$

So

$$\begin{aligned} \widetilde{B}_1 = F_{01}\check{B}_1 F_{02}^{-1} &= F_{01} \begin{pmatrix} \mathbf{b}_1^t F_{021}^t + i\mathbf{b}_1^t F_{022}^t & i(\mathbf{b}_1^t F_{021}^t + i\mathbf{b}_1^t F_{022}^t) \\ \hat{\mathbf{b}}_1^t F_{021}^t + i\hat{\mathbf{b}}_1^t F_{022}^t & i(\hat{\mathbf{b}}_1^t F_{021}^t + i\hat{\mathbf{b}}_1^t F_{022}^t) \end{pmatrix} \\ &= \begin{pmatrix} b_1^t & ib_1^t \\ \hat{b}_1^t & i\hat{b}_1^t \end{pmatrix}. \end{aligned}$$

By a conjugation of (see Lemma 4.14)

$$Q = \begin{pmatrix} I_2 & 0 \\ 0 & Q_2 \end{pmatrix} \quad \text{with } Q_2 = \begin{pmatrix} 1 & 0 & & & & & & \\ & & 1 & 0 & & & & \\ & & & & \cdots & & & \\ & & & & & & 1 & 0 \\ 0 & 1 & & & & & & \\ & & 0 & 1 & & & & \\ & & & & \cdots & & & \\ & & & & & & 0 & 1 \end{pmatrix},$$

we see that the normalized potential $\eta = Q^{-1}\tilde{\eta}Q$ has the desired form (4.11). □

The converse part of Theorem 4.21 needs a detailed discussion of the Iwasawa decompositions of F_- (see the next section).

4.5 Potentials corresponding to H-totally isotropic surfaces

In this section we will first give a characterization of H-totally isotropic surfaces in terms of normalized potentials. This also provides a procedure to construct examples. As illustrations we derive two kinds of examples. We will state the main results in this section and leave the computations to the next section.

4.5.1 The characterization of H-totally isotropic surfaces

Theorem 4.22. *Let η be a normalized potential of the form* (4.11)*. Let $\mathcal{F} = Y \wedge \hat{Y}$ be the corresponding isotropic harmonic map. Then $[Y]$ and $[\hat{Y}]$ are a pair of $H-$totally isotropic adjoint Willmore surfaces on the points they are immersed. And $\mathcal{F}$ is a harmonic map of finite uniton number at most* 2*.*

To prove Theorem 4.22, one need to perform an Iwasawa decomposition. A simple way to do this is to make the potential in (4.11) being of strictly upper-triangle matrices. For this purpose, we will need a Lie group isometry. Then under this isometry, we can write down the Iwasawa decomposition in an explicit way. As a consequence, we can derive some geometric properties of the corresponding Willmore surfaces.

First we define a new Lie group as below

$$G(n,\mathbb{C}) = \{A \in Mat(n,\mathbb{C}) | A^t J_n A = J_n, \det A = 1\},$$

with

$$J_n = \begin{pmatrix} & & 1 \\ & \cdot^{\cdot^{\cdot}} & \\ 1 & & \end{pmatrix}.$$

Theorem 4.22 can be derived from the following lemmas.

Lemma 4.23. *We have the Lie group isometry*

$$\begin{array}{lll} \mathcal{P}: SO(1,2m+1,\mathbb{C}) & \rightarrow G(2m+2,\mathbb{C}) \\ \qquad A & \mapsto \tilde{P}_1^{-1}\tilde{P}^{-1}A\tilde{P}\tilde{P}_1, \end{array} \tag{4.13}$$

with

$$\tilde{P} = \frac{1}{\sqrt{2}} \begin{pmatrix} 1 & & & & & & & -1 \\ 1 & & & & & & & 1 \\ & -i & & & & & i & \\ & 1 & & & & & 1 & \\ & & \ddots & & & \iddots & & \\ & & & -i & i & & & \\ & & & 1 & 1 & & & \end{pmatrix} \quad \text{and}$$

$$\tilde{P}_1 = \begin{pmatrix} 0 & 1 & 0 & 0 \\ I_m & 0 & 0 & 0 \\ 0 & 0 & 0 & I_m \\ 0 & 0 & 1 & 0 \end{pmatrix}.$$

Moreover, we have the following results.

1. $\mathcal{P}(SO(1,2m+1)) = \left\{B \in G(2m+2,\mathbb{C}) | B = S_0 \bar{B} S_0^{-1}\right\}$ *with*

   $$S_0 = \tilde{P}_1^{-1}\tilde{P}^{-1}\bar{\tilde{P}}\bar{\tilde{P}}_1 = \begin{pmatrix} 0 & 0 & J_m \\ 0 & I_2 & 0 \\ J_m & 0 & 0 \end{pmatrix}.$$

   So $\mathcal{P}(\Lambda SO(1,2m+1)) = \{F \in \Lambda G(2m+2,\mathbb{C}) | \tau(F) = F\}$ *with*

   $$\begin{array}{rl} \tau : G(2m+2,\mathbb{C}) & \to G(2m+2,\mathbb{C}) \\ F & \mapsto S_0 \bar{F} S_0^{-1}. \end{array}$$

   And $\tau(F)^{-1} = \hat{J}\bar{F}^t\hat{J}^{-1}$ *with*

   $$\hat{J} = \begin{pmatrix} I_m & & \\ & J_2 & \\ & & I_m \end{pmatrix}.$$

2. *The image of the subgroup* $K^{\mathbb{C}} = SO(1,1,\mathbb{C}) \times SO(2m,\mathbb{C})$ *in* $G(2m+2,\mathbb{C})$ *is*

   $$\mathcal{P}(K^{\mathbb{C}}) = \{B \in G(2m+2,\mathbb{C}) |\ B = D_0 B D_0^{-1}\},$$

   with $D_0 = \tilde{P}_1^{-1}\tilde{P}^{-1}D\tilde{P}\tilde{P}_1 = diag\,(I_m, -I_2, I_m)$, $D = diag\,(-I_2, I_{2m})$.

Lemma 4.24. *Under the isometry of* (4.13), *we have the following results*

1. *For* η_{-1} *in* (4.11), *one has*

   $$\mathcal{P}(\eta_{-1}) = \begin{pmatrix} 0 & \check{f} & 0 \\ 0 & 0 & -J_m\check{f}^t J_2 \\ 0 & 0 & 0 \end{pmatrix}, \tag{4.14}$$

with

$$\check{f} = \begin{pmatrix} \check{f}_{11} & \check{f}_{12} \\ \vdots & \vdots \\ \check{f}_{m1} & \check{f}_{m2} \end{pmatrix}, \ \check{f}^{\sharp} := J_m \check{f}^t J_2,$$

and

$$\check{f}_{j1} = i(h_{j1} - \hat{h}_{j1}), \ \check{f}_{j2} = -i(h_{j1} + \hat{h}_{j1}), \ j = 1, \cdots, m.$$

2. *Let H be a solution to $H^{-1}dH = \lambda^{-1}\mathcal{P}(\eta_{-1})\mathrm{d}z$, $H(0,0,\lambda) = I$. Then*

$$H = I + \lambda^{-1}H_1 + \lambda^{-2}H_2 = \begin{pmatrix} I & \lambda^{-1}f & \lambda^{-2}g \\ 0 & I & -\lambda^{-1}f^{\sharp} \\ 0 & 0 & I \end{pmatrix},$$

with $f^{\sharp} := J_m f^t J_2$, $f = \int_0^z \check{f}dz$ and $g = -\int_0^z f\check{f}^{\sharp}dz$.

3. *We have the Iwasawa decomposition of H as follows*

$$\begin{aligned} \tilde{F} &= H\tau(W)L^{-1} \\ &= \begin{pmatrix} (I - f\bar{u}^{\sharp} - gJ\bar{v}J)l_1^{-1} & \lambda^{-1}(f + gJ\bar{u})l_0^{-1} & \lambda^{-2}gl_4^{-1} \\ -\lambda(\bar{u}^{\sharp}J + f^{\sharp}J\bar{v}J)l_1^{-1} & (I - f^{\sharp}J\bar{u})l_0^{-1} & -\lambda^{-1}f^{\sharp}l_4^{-1} \\ \lambda^2 J\bar{v}Jl_1^{-1} & \lambda J\bar{u}l_0^{-1} & l_4^{-1} \end{pmatrix}. \end{aligned} \tag{4.15}$$

Here $W = I + \lambda^{-1}W_1 + \lambda^{-2}W_2$ and $L = diag\{l_1, l_0, l_4\}$ satisfy

$$W_0 = \begin{pmatrix} a & 0 & 0 \\ 0 & q & 0 \\ 0 & 0 & \varrho \end{pmatrix} = \tau(L)^{-1}L, \ W_1 = \begin{pmatrix} 0 & u & 0 \\ 0 & 0 & -u^{\sharp} \\ 0 & 0 & 0 \end{pmatrix},$$

$$W_2 = \begin{pmatrix} 0 & 0 & v \\ 0 & 0 & 0 \\ 0 & 0 & 0 \end{pmatrix}.$$

Here a, q, d, u, v are solutions to the following equation

$$a + uqJ\bar{u}^t + v\varrho\bar{v}^t = I, \tag{4.16a}$$

$$uq - v\varrho\bar{u}^{\sharp t}J = f, \tag{4.16b}$$

$$v\varrho = g, \tag{4.16c}$$

$$q + u^{\sharp}\varrho\bar{u}^{\sharp t}J = I + J\bar{f}^t f, \tag{4.16d}$$

$$u^{\sharp}\varrho = f^{\sharp} - J\bar{f}^t g, \tag{4.16e}$$

$$\varrho = I + \bar{f}^{t\sharp}Jf^{\sharp} + \bar{g}^t g. \tag{4.16f}$$

4. *The M-C form of $\tilde{F}$ has the form*

$$\tilde{\alpha}_1' = \begin{pmatrix} 0 & l_1 f' l_0^{-1} & 0 \\ 0 & 0 & -l_0 f^{\sharp\prime} l_4^{-1} \\ 0 & 0 & 0 \end{pmatrix} \mathrm{d}z, \tag{4.17}$$

$$\tilde{\alpha}_0' =$$

$$\lambda^{-1}\begin{pmatrix} -f'\bar{u}^{\sharp}J - l_{1z}l_1^{-1} & 0 & 0 \\ 0 & -\bar{u}^{\sharp}Jf' - f^{\sharp\prime}J\bar{u} - l_{0z}l_0^{-1} & 0 \\ 0 & 0 & -J\bar{u}f^{\sharp\prime} - l_{4z}l_4^{-1} \end{pmatrix}\mathrm{d}z. \tag{4.18}$$

Lemma 4.25. *Under the isometry of* (4.13), y *is an* $H-$*totally isotropic surface with an adjoint transform* $\hat{y} = [\hat{Y}]$.

Lemma 4.26. *Under the isometry of* (4.13), $\mathcal{P}(\eta_{-1})$ *takes value in the nilpotent Lie sub-algebra*

$$\mathfrak{g}_{nil} = \left\{ X \in \Lambda\mathfrak{g}(2m+2,\mathbb{C}) \,\middle|\, X = \lambda^{-1}\begin{pmatrix} 0 & \cdots & 0 \\ 0 & 0 & \cdots \\ 0 & 0 & 0 \end{pmatrix} + \lambda^{-2}\begin{pmatrix} 0 & 0 & \cdots \\ 0 & 0 & 0 \\ 0 & 0 & 0 \end{pmatrix} \right\}.$$

As a consequence, $\mathcal{F} = Y \wedge \hat{Y}$ *is of finite uniton number at most* 2.

Here the proof of Lemma 4.23 is a straightforward computation so that we leave it to the interested readers. Lemma 4.26 holds apparently. The proof of Lemma 4.24 and 4.25 will be given in Section 6, since it involves many technical computations.

4.5.2 Constructions of examples

In this subsection we will provide two kinds of examples. The first one concerns the new Willmore two-sphere derived in [9], which is the first example of Willmore two-spheres in S^6 admitting no dual surfaces. Here we will derive this surface together with one of its adjoint surface. This also indicates that it can be derived from an adjoint transform of some minimal surface in $\mathbb{R}^6$.

The next example concerns one of the most simple Willmore two-spheres in S^4, i.e., the one derived by a holomorphic curve in $\mathbb{C}^2 = \mathbb{R}^4$ with total curvature -4π.

Theorem 4.27. *Let*

$$\eta = \lambda^{-1}\eta_{-1}\mathrm{d}z \quad \textit{with } \hat{B}_1 = \frac{1}{2}\begin{pmatrix} -i & 1 & i & -1 & -2iz & 2z \\ i & -1 & i & -1 & 2iz & -2z \end{pmatrix}.$$

Then the associated family of the corresponding isotropic harmonic maps is

$Y\wedge\hat{Y}$, with

$$Y=-\frac{\sqrt{2}}{2\varsigma}\begin{pmatrix} -1-r^2-\frac{r^4}{4}-\frac{r^6}{9}\\ 1-r^2+\frac{r^4}{4}+\frac{r^6}{9}\\ \frac{ir^2}{2}(\lambda^{-1}z-\lambda\bar{z})\\ -\frac{r^2}{2}(\lambda^{-1}z+\lambda\bar{z})\\ -i(\lambda^{-1}z-\lambda\bar{z})\\ \lambda^{-1}z+\lambda\bar{z}\\ \frac{ir^2}{3}(\lambda^{-1}z^2-\lambda\bar{z}^2)\\ -\frac{r^2}{3}(\lambda^{-1}z^2+\lambda\bar{z}^2)\end{pmatrix}, \tag{4.19}$$

$$\hat{Y}=\frac{\sqrt{2}}{2\varsigma}\begin{pmatrix} 1+r^2+\frac{5r^4}{4}+\frac{4r^6}{9}+\frac{r^8}{36}\\ 1-r^2-\frac{3r^4}{4}+\frac{4r^6}{9}-\frac{r^8}{36}\\ -i(\lambda^{-1}z-\lambda\bar{z})(1+\frac{r^6}{9})\\ (\lambda^{-1}z+\lambda\bar{z})(1+\frac{r^6}{9})\\ \frac{ir^2}{2}(\lambda^{-1}z-\lambda\bar{z})(1+\frac{4r^2}{3})\\ -\frac{r^2}{2}(\lambda^{-1}z+\lambda\bar{z})(1+\frac{4r^2}{3})\\ -i(\lambda^{-1}z^2-\lambda\bar{z}^2)(1-\frac{r^4}{12})\\ (\lambda^{-1}z^2+\lambda\bar{z}^2)(1-\frac{r^4}{12})\end{pmatrix}. \tag{4.20}$$

Here $r=|z|$, $\varsigma=\left|1-\frac{r^4}{4}-\frac{2r^6}{9}\right|$. Moreover, we have

1. *Set $\hat{Y}=(\hat{y}_0,\cdots,\hat{y}_7)^t$. Then $\hat{y}=\frac{1}{\hat{y}_0}(\hat{y}_1,\cdots,\hat{y}_7)^t=[\hat{Y}]$ is an H-totally isotropic, Willmore immersion from S^2 to S^6, with metric*

$$|\hat{y}_z|^2|\mathrm{d}z|^2=\frac{2(1+4r^2+\frac{r^4}{4}+\frac{2r^6}{9}+\frac{4r^8}{9}+\frac{r^{10}}{36}+\frac{r^{12}}{81})}{\left(1+r^2+\frac{5r^4}{4}+\frac{4r^6}{9}+\frac{r^8}{36}\right)^2}|\mathrm{d}z|^2.$$

   $[\hat{Y}]$ has no dual surface.

2. *Set $Y=(y_0,\cdots,y_7)^t$. Then $y=\frac{1}{y_0}(y_1,\cdots,y_7)^t=[Y]$ is an H-totally isotropic, Willmore immersion from $\mathbb{C}$ to S^6, with metric*

$$|y_z|^2|\mathrm{d}z|^2=\frac{2(1+\frac{r^4}{4}+\frac{4r^6}{9})}{\left(1+r^2+\frac{r^4}{4}+\frac{r^6}{9}\right)^2}|\mathrm{d}z|^2.$$

   Note that $[Y]$ is a Willmore map from S^2, with a branched point $z=\infty$. Moreover, y is conformally equivalent to the minimal surface x in $\mathbb{R}^8$:

$$x=\begin{pmatrix} i(\lambda^{-1}z-\lambda\bar{z})\\ -(\lambda^{-1}z+\lambda\bar{z})\\ -\frac{2i}{r^2}\left(\lambda^{-1}z-\lambda\bar{z}\right)\\ \frac{2}{r^2}\left(\lambda^{-1}z+\lambda\bar{z}\right)\\ \frac{2i}{3}(\lambda^{-1}z^2-\lambda\bar{z}^2)\\ -\frac{2}{3}(\lambda^{-1}z^2+\lambda\bar{z}^2)\end{pmatrix}.$$

3. *The harmonic map* $Y\wedge\hat{Y}$ *has no definition on the curve* $1-\frac{r^4}{4}-\frac{2r^6}{9}=0$. *But the maps* $[Y]$ *and* $[\hat{Y}]$ *are well defined on the whole two-sphere* S^2.

Remark 5.

1. *From this we see that it is possible that although the harmonic map* $Y\wedge\hat{Y}$ *is not globally well-defined, the Willmore surfaces* $[Y]$ *and* $[\hat{Y}]$ *are well-defined. This is a very interesting phenomenon to be explained, which may be related to the Iwasawa decompositions of the loop group* $\Lambda G^{\mathbb{C}}_{\sigma}$ *of the non-compact group* $G=SO^+(1,2m+1)$.

2. *One can also derive* $\hat{y}$ *from a concrete adjoint transform of the minimal surface* x. *To ensure* $\hat{y}$ *to be immersed, one needs some restrictions on the minimal surface* x. *Our examples here play an important role in the discussions of these conditions. We refer to [17] for more details.*

Theorem 4.28. *Let*

$$\eta=\lambda^{-1}\eta_{-1}\mathrm{d}z \quad \textit{with } \hat{B}_1=\frac{1}{2}\begin{pmatrix} i & -1 & -i & 1\\ i & -1 & i & -1\end{pmatrix}.$$

Then the associated family of the corresponding isotropic harmonic maps is $Y\wedge\hat{Y}$, *with*

$$Y=\frac{\sqrt{2}}{2\varsigma}\begin{pmatrix}(1+\frac{r^2}{2})^2\\ -(1-\frac{r^2}{2})^2\\ i(\lambda^{-1}z-\lambda\bar{z})\\ -\lambda^{-1}z-\lambda\bar{z}\\ -\frac{ir^2}{2}(\lambda^{-1}z-\lambda\bar{z})\\ \frac{r^2}{2}(\lambda^{-1}z+\lambda\bar{z})\end{pmatrix},\quad \hat{Y}=\frac{\sqrt{2}}{2\varsigma}\begin{pmatrix}(1+\frac{r^2}{2})^2\\ (1-\frac{r^2}{2})^2\\ \frac{ir^2}{2}(\lambda^{-1}z-\lambda\bar{z})\\ -\frac{r^2}{2}(\lambda^{-1}z+\lambda\bar{z})\\ -i(\lambda^{-1}z-\lambda\bar{z})\\ \lambda^{-1}z+\lambda\bar{z}\end{pmatrix}. \tag{4.21}$$

Here $r=|z|$ *and* $\varsigma=1-\frac{r^4}{4}$. *Note that in this case* Y *is conformally equivalent to* $\hat{Y}$. *Moreover,* Y *and* $\hat{Y}$ *satisfy the following results.*

1. *Set* $Y=(y_0,\cdots,y_5)^t$. *Then* $y=\frac{1}{y_0}(y_1,\cdots,y_5)^t=[Y]$ *is an H-totally isotropic, Willmore immersion from* S^2 *to* S^4, *with metric* $\langle y_z,y_{\bar{z}}\rangle|\mathrm{d}z|^2=\frac{2+\frac{r^4}{2}}{(1+\frac{r^2}{2})^4}|\mathrm{d}z|^2$.

2. $[Y]$ *is conformally equivalent to the minimal surface* x *in* $\mathbb{R}^4$:

$$x=\begin{pmatrix}\frac{2i\lambda^{-1}}{\bar{z}}-\frac{2i\lambda}{z} & -\frac{2\lambda^{-1}}{\bar{z}}-\frac{2\lambda}{z} & -i\lambda^{-1}z+i\lambda\bar{z} & \lambda^{-1}z+\lambda\bar{z}\end{pmatrix}^t.$$

3. *The harmonic map* $Y\wedge\hat{Y}$ *has no definition on the curve* $1-\frac{r^4}{4}=0$. *But both* $[Y]$ *and* $[\hat{Y}]$ *are Willmore immersions on the whole two-sphere* S^2.

Remark 6. *Set $\lambda = 1$. On the curve $\Gamma: 1-\frac{r^4}{4}=0$ we have $z=\sqrt{2}e^{i\theta}$ and hence*

$$(1-\frac{r^4}{4})Y|_\Gamma = \begin{pmatrix} 2\sqrt{2} & 0 & -2\sin\theta & -2\cos\theta & 2\sin\theta & 2\cos\theta \end{pmatrix},$$
$$(1-\frac{r^4}{4})\hat{Y}|_\Gamma = \begin{pmatrix} 2\sqrt{2} & 0 & -2\sin\theta & -2\cos\theta & 2\sin\theta & 2\cos\theta \end{pmatrix}.$$

So $Y\wedge\hat{Y}$ has no definition on the curve Γ.

4.6 Appendix: Iwasawa decompositions and computations of examples

This section contains two parts: the proof of the technical lemmas in Section 5.1 and the computations of the examples in Section 5.2.

4.6.1 On the technical lemmas of Section 5.1

To begin with, it is convenient to have the explicit expressions of $\mathcal{P}(A)$ in (4.13). So we will first give this expression and then provide the proofs of Lemma 4.24 and Lemma 4.25.

4.6.1.1 On $\mathcal{P}(A)$

Set

$$A=(\mathbf{a}_{ij}),\ \mathcal{P}(A)=B=(\mathbf{b}_{ij}),\ \hat{j}=2m+3-j \ \text{ and } \hat{k}=2m+3-k.$$

Then when $j=1,\cdots,m$, we have

$$\mathbf{b}_{jk}=\begin{cases} \dfrac{\mathbf{a}_{2j+1,2k+1}-i\mathbf{a}_{2j+2,2k+1}+i\mathbf{a}_{2j+1,2k+2}+\mathbf{a}_{2j+2,2k+2}}{2}, & k=1,\cdots,m;\\ \dfrac{i\mathbf{a}_{2j+1,1}+\mathbf{a}_{2j+2,1}+i\mathbf{a}_{2j+1,2}+\mathbf{a}_{2j+2,2}}{2}, & k=m+1;\\ \dfrac{-i\mathbf{a}_{2j+1,1}-\mathbf{a}_{2j+2,1}+i\mathbf{a}_{2j+1,2}+\mathbf{a}_{2j+2,2}}{2}, & k=m+2;\\ \dfrac{-\mathbf{a}_{2j+1,2\hat{k}+1}+i\mathbf{a}_{2j+2,2\hat{k}+1}+i\mathbf{a}_{2j+1,2\hat{k}+2}+\mathbf{a}_{2j+2,2\hat{k}+2}}{2}, \\ \quad k=m+3,\cdots,2m+2. \end{cases} \tag{4.22}$$

When $j = m+1$ we have

$$\mathbf{b}_{jk} = \begin{cases} \dfrac{-i\mathbf{a}_{1,2k+1} - i\mathbf{a}_{2,2k+1} + \mathbf{a}_{1,2k+2} + \mathbf{a}_{2,2k+2}}{2}, & k = 1, \cdots, m; \\ \dfrac{\mathbf{a}_{11} + \mathbf{a}_{21} + \mathbf{a}_{12} + \mathbf{a}_{22}}{2}, & k = m+1; \\ \dfrac{-\mathbf{a}_{11} - \mathbf{a}_{21} + \mathbf{a}_{12} + \mathbf{a}_{22}}{2}, & k = m+2; \\ \dfrac{i\mathbf{a}_{1,2\hat{k}+1} + i\mathbf{a}_{2,2\hat{k}+1} + \mathbf{a}_{1,2\hat{k}+2} + \mathbf{a}_{2,2\hat{k}+2}}{2}, & k = m+3, \cdots, 2m+2. \end{cases}$$

When $j = m+2$ we have

$$\mathbf{b}_{jk} = \begin{cases} \dfrac{i\mathbf{a}_{1,2k+1} - i\mathbf{a}_{2,2k+1} - \mathbf{a}_{1,2k+2} + \mathbf{a}_{2,2k+2}}{2}, & k = 1, \cdots, m; \\ \dfrac{-\mathbf{a}_{11} + \mathbf{a}_{21} - \mathbf{a}_{12} + \mathbf{a}_{22}}{2}, & k = m+1; \\ \dfrac{\mathbf{a}_{11} - \mathbf{a}_{21} - \mathbf{a}_{12} + \mathbf{a}_{22}}{2}, & k = m+2; \\ \dfrac{-i\mathbf{a}_{1,2\hat{k}+1} + i\mathbf{a}_{2,2\hat{k}+1} - \mathbf{a}_{1,2\hat{k}+2} + \mathbf{a}_{2,2\hat{k}+2}}{2}, \\ k = m+3, \cdots, 2m+2. \end{cases}$$

When $j = m+3, \cdots, 2m+2$, we have

$$\mathbf{b}_{jk} = \begin{cases} \dfrac{-\mathbf{a}_{2\hat{j}+1,2k+1} - i\mathbf{a}_{2\hat{j}+2,2k+1} - i\mathbf{a}_{2\hat{j}+1,2k+2} + \mathbf{a}_{2\hat{j}+2,2k+2}}{2}, \\ \qquad k = 1, \cdots, m; \\ \dfrac{-i\mathbf{a}_{2\hat{j}+1,1} + \mathbf{a}_{2\hat{j}+2,1} - i\mathbf{a}_{2\hat{j}+1,2} + \mathbf{a}_{2\hat{j}+2,2}}{2}, & k = m+1; \\ \dfrac{i\mathbf{a}_{2\hat{j}+1,1} - \mathbf{a}_{2\hat{j}+2,1} - i\mathbf{a}_{2\hat{j}+1,2} + \mathbf{a}_{2\hat{j}+2,2}}{2}, & k = m+2; \\ \dfrac{\mathbf{a}_{2\hat{j}+1,2\hat{k}+1} + i\mathbf{a}_{2\hat{j}+2,2\hat{k}+1} - i\mathbf{a}_{2\hat{j}+1,2\hat{k}+2} + \mathbf{a}_{2\hat{j}+2,2\hat{k}+2}}{2}, \\ \qquad k = m+3, \cdots, 2m+2. \end{cases} \tag{4.23}$$

4.6.1.2 Proof of Lemma 4.24

(1). Assume that $\eta_{-1} = (\mathrm{a}_{jk})$. Then one has

$$\mathbf{a}_{1,2j+1} = \mathbf{a}_{2j+1,1} = h_{j1}, \ \mathbf{a}_{1,2j+2} = \mathbf{a}_{2j+2,1} = ih_{j1},$$

$$\mathbf{a}_{2,2j+1} = -\mathbf{a}_{2j+1,2} = \hat{h}_{j1}, \ \mathbf{a}_{2,2j+2} = -\mathbf{a}_{2j+2,2} = i\hat{h}_{j1},$$

when $j = 1, \cdots, m+1$, and $\mathbf{a}_{jk} = 0$ otherwise. Substituting into (4.22)–(4.23), one obtains (4.14).

(2). First by definition, $H(0,0,\lambda)=I$. Next,

$$H_z=\lambda^{-1}H_{1z}+\lambda^{-2}H_{2z}=\begin{pmatrix}0&\lambda^{-1}\check{f}&-\lambda^{-2}f\check{f}^{\sharp}\\0&0&-\lambda^{-1}\check{f}^{\sharp}\\0&0&0\end{pmatrix}=H\mathcal{P}(\eta_{-1}).$$

(3). First since $H(0,0,\lambda)=I$, when $|z|<\varepsilon$, there exists an Iwasawa decomposition $H=\tilde{F}V_+$, with $\tilde{F}\in\Lambda G_\sigma$, $V_+\in\Lambda^+G_\sigma^{\mathbb{C}}$. Next we want to express $\tilde{F}$ in terms of H. Since $H=I+\lambda^{-1}H_1+\lambda^{-2}H_2$, by the reality condition we see that $V_+=V_0+\lambda V_1+\lambda^2V_2$ with V_0, V_1 and V_2 independent of λ.

Assume that $V_+=V_0\hat{V}_+$ such that $\hat{V}_+|_{\lambda=0}=I$. Then we have

$$\tilde{F}=H\hat{V}_+^{-1}V_0^{-1}.$$

Since $\tau(\tilde{F})=\tilde{F}$, we obtain $\tau(H)\tau(\hat{V}_+^{-1})\tau(V_0^{-1})=H\hat{V}_+^{-1}V_0^{-1}$, i.e.,

$$\tau(H)^{-1}H=\tau(\hat{V}_+^{-1})\tau(V_0^{-1})V_0\hat{V}_+.$$

We then assume that

$$\tau(H)^{-1}H=WW_0\tau(W)^{-1}$$

with $W_0=\tau(V_0^{-1})V_0$, $W=I+\lambda^{-1}W_1+\lambda^{-2}W_2=\tau(\hat{V}_+^{-1})$ and

$$W_0=\begin{pmatrix}a&0&b\\0&q&0\\c&0&\varrho\end{pmatrix},\quad W_1=\begin{pmatrix}0&u&0\\-u_0^{\sharp}&0&-u^{\sharp}\\0&u_0&0\end{pmatrix}.$$

Comparing the coefficients of λ, we obtain

$$\begin{cases}W_2W_0=H_2,\\W_1W_0+W_2W_0\tau(W_1)=H_1+\tau(H_1)H_2,\\W_0+W_1W_0\tau(W_1)+W_2W_0\tau(W_2)=I+\tau(H_1)H_1+\tau(H_2)H_2.\end{cases}\tag{4.24}$$

Direct computation shows that

$$\tau(H)^{-1}=\begin{pmatrix}I&0&0\\\lambda J\bar{f}^t&I&0\\\lambda^2J\bar{g}^t&-\lambda\bar{f}^{\sharp,t}J&I\end{pmatrix},$$

and

$$\tau(H)^{-1}H=\begin{pmatrix}I&\lambda^{-1}f&\lambda^{-2}g\\\lambda J\bar{f}^t&I+J\bar{f}^tf&\lambda^{-1}(J\bar{f}^t-f^{\sharp})\\\lambda^2J\bar{g}^t&-\lambda(\bar{f}^{\sharp,t}J-\bar{g}^tf)&I+\bar{f}^{\sharp,t}Jf^{\sharp}+\bar{g}^tg\end{pmatrix}.\tag{4.25}$$

From the first two equations of (4.24) we can see that

$$W_1W_0 = H_1 + \tau(H_1)H_2 - W_2W_0\tau(W_1) = H_1 + \tau(H_1)H_2 - H_2\tau(W_1)$$
$$= \begin{pmatrix} 0 & \cdots & 0 \\ 0 & 0 & \cdots \\ 0 & 0 & 0 \end{pmatrix} + \begin{pmatrix} 0 & 0 & 0 \\ 0 & 0 & \cdots \\ 0 & 0 & 0 \end{pmatrix} - \begin{pmatrix} 0 & \cdots & 0 \\ 0 & 0 & 0 \\ 0 & 0 & 0 \end{pmatrix}.$$

So

$$W_1 = \begin{pmatrix} 0 & \cdots & 0 \\ \cdots & 0 & \cdots \\ 0 & 0 & 0 \end{pmatrix}, \text{ i.e., } u_0 = 0.$$

Then from the last equation of (4.24) we see that

$$W_0 = (I + \tau(H_1)H_1 + \tau(H_2)H_2) - W_1W_0\tau(W_1) - W_2W_0\tau(W_2)$$
$$= \begin{pmatrix} \cdots & 0 & 0 \\ 0 & \cdots & 0 \\ 0 & 0 & \cdots \end{pmatrix} - \begin{pmatrix} \cdots & 0 & 0 \\ 0 & \cdots & 0 \\ 0 & 0 & 0 \end{pmatrix} - \begin{pmatrix} \cdots & 0 & 0 \\ 0 & 0 & 0 \\ 0 & 0 & 0 \end{pmatrix},$$

i.e., $b = c = 0$. So we have that

$$W_0 = \begin{pmatrix} a & 0 & 0 \\ 0 & q & 0 \\ 0 & 0 & \varrho \end{pmatrix}, \quad W = \begin{pmatrix} I & \lambda^{-1}u & \lambda^{-2}v \\ 0 & I & -\lambda^{-1}u^\sharp \\ 0 & 0 & I \end{pmatrix}$$

and

$$\tau(W)^{-1} = \begin{pmatrix} I & 0 & 0 \\ \lambda J\bar{u}^t & I & 0 \\ \lambda^2 J\bar{v}^t & -\lambda\bar{u}^{\sharp,t}J & I \end{pmatrix}.$$

So

$$WW_0\tau(W)^{-1} = \begin{pmatrix} a + uqJ\bar{u}^t + v\varrho\bar{v}^t & \lambda^{-1}(uq - v\varrho\bar{u}^{\sharp,t}J) & \lambda^{-2}v\varrho \\ \lambda(qJ\bar{u}^t - u^\sharp\varrho\bar{v}^t) & q + u^\sharp\varrho\bar{u}^{\sharp,t}J & -\lambda^{-1}u^\sharp\varrho \\ \lambda^2\varrho\bar{v}^t & -\lambda\varrho\bar{u}^{\sharp,t} & \varrho \end{pmatrix}. \tag{4.26}$$

By (4.25), (4.26) and $\tau(H)^{-1}H = WW_0\tau(W)^{-1}$ we obtain (4.16).

Apparently W_0 has the decomposition

$$W_0 = \tau(L)^{-1}L, \quad \text{with } L = \mathrm{diag}\{l_1, l_0, l_4\}.$$

Now set $\tilde{F} = H\tau(W)L^{-1}$. We see that $\tau(\tilde{F}) = \tilde{F}$ and $H = \tilde{F}L\tau(W)^{-1}$ is an Iwasawa decomposition of H. Substituting

$$\tau(W) = \begin{pmatrix} I & 0 & 0 \\ -\lambda\bar{u}^\sharp J & I & 0 \\ \lambda^2 J\bar{v}J & \lambda J\bar{u} & I \end{pmatrix},$$

L^{-1} and H into $\tilde{F} = H\tau(W)L^{-1}$, one obtains (4.15).

(4). Since

$$\tau(W) = I + \lambda\tau(W_1) + \lambda^2\tau(W_2),\ \tau(W)^{-1} = I - \lambda\tau(W_1) + \lambda^2(\cdots).$$

So

$$\tilde{\alpha}_1' = L\mathcal{P}(\eta_{-1})L^{-1}\mathrm{d}z, \quad \tilde{\alpha}_0 = \left(-\tau(W_1)\mathcal{P}(\eta_{-1}) + \mathcal{P}(\eta_{-1})\tau(W_1) - L_zL^{-1}\right)\mathrm{d}z.$$

Then (4.17) and (4.18) follow. This finishes the proof of Lemma 4.24.

4.6.1.3 Proof of Lemma 4.25

Assume that

$$\tilde{F} = \mathcal{P}(F) \text{ and } \alpha' = F^{-1}F_z\mathrm{d}z = (\mathbf{b}_{jk})\mathrm{d}z.$$

So we have $\mathcal{P}(\alpha') = \tilde{\alpha}_1' + \tilde{\alpha}_0'$. Applying (4.22)–(4.23), (4.17) and (4.18), we obtain that

$$\frac{-\mathbf{b}_{2j+1,2\hat{k}+1} + i\mathbf{b}_{2j+2,2\hat{k}+1} + i\mathbf{b}_{2j+1,2\hat{k}+2} + \mathbf{b}_{2j+2,2\hat{k}+2}}{2} = 0,$$

when $1 \le j \le m,\ m+3 \le k \le 2m+2$;

$$\frac{-i\mathbf{b}_{1,2k+1} - i\mathbf{b}_{2,2k+1} + \mathbf{b}_{1,2k+2} + \mathbf{b}_{2,2k+2}}{2} = 0, \quad \text{when } 1 \le k \le m;$$

$$\frac{i\mathbf{b}_{1,2k+1} - i\mathbf{b}_{2,2k+1} - \mathbf{b}_{1,2k+2} + \mathbf{b}_{2,2k+2}}{2} = 0, \quad \text{when } \le k \le m;$$

$$\frac{-\mathbf{b}_{2\hat{j}+1,2k+1} - i\mathbf{b}_{2\hat{j}+2,2k+1} - i\mathbf{b}_{2\hat{j}+1,2k+2} + \mathbf{b}_{2\hat{j}+2,2k+2}}{2} = 0,$$

when $m+3 \le j \le 2m+2, 1 \le k \le m$. Here $\hat{j} = 2m+3-j$, $\hat{k} = 2m+3-k$. Since $\mathbf{b}_{jk} = -\mathbf{b}_{kj}$ for $j,k > 1$, and $\mathbf{b}_{1k} = \mathbf{b}_{k1}$ for $k > 1$, we have

$$\begin{cases} \mathbf{b}_{j,2k+2} = i\mathbf{b}_{j,2k+1}, & j = 1,2, \quad 1 \le k \le m; \\ \mathbf{b}_{2j+1,2k+1} = \mathbf{b}_{2j+2,2k+2}, & 1 \le j,k \le m; \\ \mathbf{b}_{2j+1,2k+2} = -\mathbf{b}_{2j+2,2k+1}, & 1 \le j,k \le m. \end{cases}$$

Set

$$F = (e_0, \hat{e}_0, \psi_1, \hat{\psi}_1, \cdots, \psi_m, \hat{\psi}_m), \quad Y = \frac{\sqrt{2}}{2}(e_0 - \hat{e}_0) \text{ and } \hat{Y} = \frac{\sqrt{2}}{2}(e_0 + \hat{e}_0).$$

We have then

$$\begin{cases} e_{0z} = \mathbf{b}_{21}\hat{e}_0 + \sum\limits_{1\le j\le m} \mathbf{b}_{1,2j+1}(\psi_j + i\hat{\psi}_j), \\ \hat{e}_{0z} = \mathbf{b}_{21}e_0 - \sum\limits_{1\le j\le m} \mathbf{b}_{2,2j+1}(\psi_j + i\hat{\psi}_j), \\ \psi_{jz} = \mathbf{b}_{1,2j+1}e_0 + \mathbf{b}_{2,2j+1}\hat{e}_0 - \sum\limits_{1\le k\le m}\left(\mathbf{b}_{2j+1,2k+1}\psi_k + \mathbf{b}_{2j+1,2k+2}\hat{\psi}_k\right), \\ \hat{\psi}_{jz} = \mathbf{b}_{1,2j+2}e_0 + \mathbf{b}_{2,2j+2}\hat{e}_0 - \sum\limits_{1\le k\le m}\left(\mathbf{b}_{2j+2,2k+1}\psi_k + \mathbf{b}_{2j+2,2k+2}\hat{\psi}_k\right). \end{cases}$$

So

$$(\psi_j + i\hat{\psi}_j)_z = -\sum_{1\le k\le m} (\mathbf{b}_{2j+1,2k+1} + i\mathbf{b}_{2j+2,2k+1})\left(\psi_k + i\hat{\psi}_k\right) \mod \{Y, \hat{Y}\}.$$

$$(\psi_j + i\hat{\psi}_j)_{\bar{z}} = -\sum_{1\le k\le m} (\overline{\mathbf{b}_{2j+1,2k+1}} + i\overline{\mathbf{b}_{2j+2,2k+1}})\left(\psi_k + i\hat{\psi}_k\right) \mod \{Y, \hat{Y}\}.$$

Since

$$Y_z = -\mathbf{b}_{21}Y + \frac{\sqrt{2}}{2}\sum_j(\mathbf{b}_{1,2j+1} + \mathbf{b}_{2,2j+1})(\psi_j + i\hat{\psi}_j)$$

and

$$\hat{Y}_z = \mathbf{b}_{21}\hat{Y} + \frac{\sqrt{2}}{2}\sum_j(\mathbf{b}_{1,2j+1} - \mathbf{b}_{2,2j+1})(\psi_j + i\hat{\psi}_j),$$

it is straightforward to verify that Y and $\hat{Y}$ satisfy (4.9) and (4.10). This finishes the proof of Lemma 4.25.

4.6.2 Computations on the examples

This subsection is to derive the examples stated in Section 5. To begin with, first we recall the formula of expressing Y and $\hat{Y}$ by elements of H. Then we will apply the formula to derive the examples.

4.6.2.1 From frame to Willmore surfaces

Suppose that $F = (e_0, \hat{e}_0, \psi_1, \hat{\psi}_1, \cdots, \psi_m, \hat{\psi}_m)$, and

$$Y = \frac{\sqrt{2}}{2}(e_0 - \hat{e}_0),\ \hat{Y} = \frac{\sqrt{2}}{2}(e_0 + \hat{e}_0)$$

Then $y = [Y]$ and $\hat{y} = [\hat{Y}]$ are the two Willmore surfaces (which may have branched points) adjoint to each other.

Assume that $F = (\mathbf{c}_{jk})$, $\tilde{F} = (\widetilde{\mathbf{c}}_{jk})$. Then by (4.22)–(4.23), we have that

$$\begin{cases} \widetilde{\mathbf{c}}_{j,m+1} + \widetilde{\mathbf{c}}_{\hat{j},m+1} = \mathbf{c}_{2j+2,1} + \mathbf{c}_{2j+2,2}, \\ -i(\widetilde{\mathbf{c}}_{j,m+1} - \widetilde{\mathbf{c}}_{\hat{j},m+1}) = \mathbf{c}_{2j+1,1} + \mathbf{c}_{2j+1,2}, \\ \widetilde{\mathbf{c}}_{m+1,m+1} + \widetilde{\mathbf{c}}_{m+2,m+1} = \mathbf{c}_{21} + \mathbf{c}_{22}, \\ \widetilde{\mathbf{c}}_{m+1,m+1} - \widetilde{\mathbf{c}}_{m+2,m+1} = \mathbf{c}_{11} + \mathbf{c}_{12}, \end{cases}$$

$$\begin{cases} \widetilde{\mathbf{c}}_{j,m+2} + \widetilde{\mathbf{c}}_{\hat{j},m+2} = -\mathbf{c}_{2j+2,1} + \mathbf{c}_{2j+2,2}, \\ -i(\widetilde{\mathbf{c}}_{j,m+2} - \widetilde{\mathbf{c}}_{\hat{j},m+2}) = -\mathbf{c}_{2j+1,1} + \mathbf{c}_{2j+1,2}, \\ \widetilde{\mathbf{c}}_{m+1,m+2} + \widetilde{\mathbf{c}}_{m+2,m+2} = -\mathbf{c}_{21} + \mathbf{c}_{22}, \\ \widetilde{\mathbf{c}}_{m+1,m+2} - \widetilde{\mathbf{c}}_{m+2,m+2} = -\mathbf{c}_{11} + \mathbf{c}_{12}. \end{cases}$$

So we have

$$Y = -\frac{\sqrt{2}}{2}\begin{pmatrix} \widetilde{\mathbf{c}}_{m+1,m+2} - \widetilde{\mathbf{c}}_{m+2,m+2} \\ \widetilde{\mathbf{c}}_{m+1,m+2} + \widetilde{\mathbf{c}}_{m+2,m+2} \\ -i(\widetilde{\mathbf{c}}_{1,m+2} - \widetilde{\mathbf{c}}_{2m+2,m+2}) \\ \widetilde{\mathbf{c}}_{1,m+2} + \widetilde{\mathbf{c}}_{2m+2,m+2} \\ \cdots \\ -i(\widetilde{\mathbf{c}}_{m,m+2} - \widetilde{\mathbf{c}}_{m+3,m+2}) \\ \widetilde{\mathbf{c}}_{m,m+2} + \widetilde{\mathbf{c}}_{m+3,m+2} \end{pmatrix},$$
$$\hat{Y} = \frac{\sqrt{2}}{2}\begin{pmatrix} \widetilde{\mathbf{c}}_{m+1,m+1} - \widetilde{\mathbf{c}}_{m+2,m+1} \\ \widetilde{\mathbf{c}}_{m+1,m+1} + \widetilde{\mathbf{c}}_{m+2,m+1} \\ -i(\widetilde{\mathbf{c}}_{1,m+1} - \widetilde{\mathbf{c}}_{2m+2,m+1}) \\ \widetilde{\mathbf{c}}_{1,m+21} + \widetilde{\mathbf{c}}_{2m+2,m+1} \\ \cdots \\ -i(\widetilde{\mathbf{c}}_{m,m+1} - \widetilde{\mathbf{c}}_{m+3,m+1}) \\ \widetilde{\mathbf{c}}_{m,m+1} + \widetilde{\mathbf{c}}_{m+3,m+1} \end{pmatrix}. \tag{4.27}$$

4.6.2.2 Proof of Theorem 4.27

Set $r = |z|$. By (4.14) we have

$$\check{f} = \begin{pmatrix} 1 & 0 \\ 0 & 1 \\ 2z & 0 \end{pmatrix}.$$

By integration one has

$$f = \begin{pmatrix} z & 0 \\ 0 & z \\ z^2 & 0 \end{pmatrix} \quad \text{and} \quad g = -\begin{pmatrix} 0 & \frac{z^2}{2} & 0 \\ \frac{2z^3}{3} & 0 & \frac{z^2}{2} \\ 0 & \frac{z^3}{3} & 0 \end{pmatrix}.$$

Substituting into (4.16f), one obtains

$$\varrho = \begin{pmatrix} 1 + \frac{4r^6}{9} & r^2\bar{z} & \frac{r^4\bar{z}}{3} \\ r^2 z & 1 + \frac{r^4}{4} + \frac{r^6}{9} & r^2 \\ \frac{r^4 z}{3} & r^2 & 1 + \frac{r^4}{4} \end{pmatrix},$$

with

$$\varrho^{-1} =$$
$$\frac{1}{|\varrho|}\begin{pmatrix} (1-\frac{r^4}{4})^2 + \frac{4r^6}{9}(1+\frac{r^4}{4}) & -\bar{z}r^2(1-\frac{r^4}{12}) & \frac{2r^4\bar{z}}{3}(1-\frac{r^4}{8}-\frac{r^6}{18}) \\ -zr^2(1-\frac{r^4}{12}) & 1 + \frac{r^4}{4} + \frac{4r^6}{9} & -r^2(1+\frac{r^6}{9}) \\ \frac{2r^4 z}{3}(1-\frac{r^4}{8}-\frac{r^6}{18}) & -r^2(1+\frac{r^6}{9}) & (1-\frac{2r^6}{9})^2 + \frac{r^4}{4}(1+\frac{4r^6}{9}) \end{pmatrix}.$$

Here $|\varrho| = \varsigma^2$ and $\varsigma = 1 - \frac{r^4}{4} - \frac{2r^6}{9}$. Then by (4.16e) one obtains

$$\begin{aligned} u^\sharp &= (f^\sharp - Jf^t g)\varrho^{-1} \\ &= \frac{z}{\varsigma}\begin{pmatrix} -\frac{r^2 z}{3} & 1 & -\frac{r^2}{2} \\ z(1-\frac{r^4}{12}) & -\frac{r^2}{2} - \frac{2r^4}{3} & 1+\frac{r^6}{9} \end{pmatrix}. \end{aligned}$$

So

$$u = \frac{z}{\varsigma}\begin{pmatrix} 1+\frac{r^6}{9} & -\frac{r^2}{2} \\ -\frac{r^2}{2} - \frac{2r^4}{3} & 1 \\ z(1-\frac{r^4}{12}) & -\frac{r^2 z}{3} \end{pmatrix}.$$

Substituting f, g and u into (4.16d), one obtains

$$q = I_2.$$

So $l_0 = I_2$. By (4.15), since

$$J\bar{u} = \frac{\bar{z}}{\varsigma}\begin{pmatrix} \bar{z}(1-\frac{r^4}{12}) & -\frac{r^2\bar{z}}{3} \\ -\frac{r^2}{2} - \frac{2r^4}{3} & 1 \\ 1+\frac{r^6}{9} & -\frac{r^2}{2} \end{pmatrix},$$

$$f + gJ\bar{u} = \frac{z}{\varsigma}\begin{pmatrix} 1+\frac{r^6}{9} & -\frac{r^2}{2} \\ -\frac{r^2}{2}(1+\frac{4r^2}{3}) & 1 \\ z(1-\frac{r^4}{12}) & -\frac{r^2 z}{3} \end{pmatrix},$$

$$I - f^\sharp J\bar{u} = \frac{1}{\varsigma}\begin{pmatrix} 1+\frac{r^4}{4}+\frac{4r^6}{9} & -r^2 \\ -r^2(1+r^2+\frac{r^6}{36}) & 1+\frac{r^4}{4}+\frac{r^6}{9} \end{pmatrix},$$

we have

$$\begin{pmatrix} \widetilde{\mathbf{c}}_{14} & \widetilde{\mathbf{c}}_{15} \\ \widetilde{\mathbf{c}}_{24} & \widetilde{\mathbf{c}}_{25} \\ \widetilde{\mathbf{c}}_{34} & \widetilde{\mathbf{c}}_{35} \\ \widetilde{\mathbf{c}}_{44} & \widetilde{\mathbf{c}}_{45} \\ \widetilde{\mathbf{c}}_{54} & \widetilde{\mathbf{c}}_{55} \\ \widetilde{\mathbf{c}}_{64} & \widetilde{\mathbf{c}}_{65} \\ \widetilde{\mathbf{c}}_{74} & \widetilde{\mathbf{c}}_{75} \\ \widetilde{\mathbf{c}}_{84} & \widetilde{\mathbf{c}}_{85} \end{pmatrix} = \frac{1}{\varsigma}\begin{pmatrix} \lambda^{-1}z(1+\frac{r^6}{9}) & -\lambda^{-1}\frac{zr^2}{2} \\ -\lambda^{-1}\frac{r^2 z}{2}(1+\frac{4r^2}{3}) & \lambda^{-1}z \\ \lambda^{-1}z^2(1-\frac{r^4}{12}) & -\lambda^{-1}\frac{r^2 z^2}{3} \\ 1+\frac{r^4}{4}+\frac{4r^6}{9} & -r^2 \\ -r^2(1+r^2+\frac{r^6}{36}) & 1+\frac{r^4}{4}+\frac{r^6}{9} \\ \lambda\bar{z}^2(1-\frac{r^4}{12}) & -\lambda\frac{r^2\bar{z}^2}{3} \\ -\lambda\frac{r^2\bar{z}}{2}(1+\frac{4r^2}{3}) & \lambda\bar{z} \\ \lambda\bar{z}(1+\frac{r^6}{9}) & -\lambda\frac{r^2\bar{z}}{2} \end{pmatrix}.$$

Substituting $\widetilde{\mathbf{c}}_{jk}$ into (4.27), one derives (4.19)–(4.20).

The rest are straightforward computations, except $\hat{y}$ being branched at $z = \infty$ and $\hat{y}$ being unbranched at $z = \infty$. To this end, we need to use another coordinate. Set $\tilde{z} = \frac{1}{z}$ and $\tilde{r} = \sqrt{|\tilde{z}|}$, we have that

$$|y_{\tilde{z}}|^2|\mathrm{d}\tilde{z}|^2 = \frac{2\tilde{r}^2(\tilde{r}^6 + \frac{\tilde{r}^2}{4} + \frac{4}{9})}{\left(\tilde{r}^6 + \tilde{r}^4 + \frac{\tilde{r}^2}{4} + \frac{1}{9}\right)^2}|\mathrm{d}\tilde{z}|^2,$$

$$|\hat{y}_{\tilde{z}}|^2|\mathrm{d}\tilde{z}|^2 = \frac{2\tilde{r}^{12} + 8\tilde{r}^{10} + \frac{\tilde{r}^8}{2} + \frac{4\tilde{r}^6}{9} + \frac{8\tilde{r}^4}{9} + \frac{\tilde{r}^2}{18} + \frac{2}{81}}{\left(\tilde{r}^8 + \tilde{r}^6 + \frac{5\tilde{r}^4}{4} + \frac{4\tilde{r}^2}{9} + \frac{1}{36}\right)^2}|\mathrm{d}\tilde{z}|^2.$$

At $\tilde{z} = 0$, $|y_{\tilde{z}}|^2|\mathrm{d}\tilde{z}|^2 = 0$ and $|\hat{y}_{\tilde{z}}|^2|\mathrm{d}\tilde{z}|^2 = 32|\mathrm{d}\tilde{z}|^2$. This finishes the proof.

4.6.2.3 Proof of Theorem 4.28

Set $r = |z|$. By (4.14) we have

$$\check{f} = \begin{pmatrix} 0 & 1 \\ 1 & 0 \end{pmatrix} \quad \Rightarrow \quad f = \begin{pmatrix} 0 & z \\ z & 0 \end{pmatrix}, \ g = -\frac{z^2}{2}\begin{pmatrix} 1 & 0 \\ 0 & 1 \end{pmatrix}.$$

Substituting into (4.16f), one obtains

$$\varrho = \begin{pmatrix} 1 + \frac{r^4}{4} & r^2 \\ r^2 & 1 + \frac{r^4}{4} \end{pmatrix} \quad \text{and} \quad \varrho^{-1} = \frac{1}{|\varrho|}\begin{pmatrix} 1 + \frac{r^4}{4} & -r^2 \\ -r^2 & 1 + \frac{r^4}{4} \end{pmatrix}.$$

Here $|\varrho| = \varsigma^2$ with $\varsigma = 1 - \frac{r^4}{4}$. Next one computes

$$u^\sharp = \frac{z}{\varsigma}\begin{pmatrix} -\frac{r^2}{2} & 1 \\ 1 & -\frac{r^2}{2} \end{pmatrix}, \ u = \frac{z}{\varsigma}\begin{pmatrix} -\frac{r^2}{2} & 1 \\ 1 & -\frac{r^2}{2} \end{pmatrix} \quad \text{and} \quad q = I_2.$$

So $l_0 = I_2$. Then we have

$$\begin{aligned} \lambda J\bar{u}l_0^{-1} &= \frac{\lambda\bar{z}}{\varsigma}\begin{pmatrix} 1 & -\frac{r^2}{2} \\ -\frac{r^2}{2} & 1 \end{pmatrix}, \\ \lambda^{-1}(f + gJ\bar{u})l_0^{-1} &= \frac{\lambda^{-1}z}{\varsigma}\begin{pmatrix} -\frac{r^2}{2} & 1 \\ 1 & -\frac{r^2}{2} \end{pmatrix} \end{aligned}$$

and

$$(I - f^\sharp J\bar{u})l_0^{-1} = \frac{1}{\varsigma}\begin{pmatrix} 1 + \frac{r^4}{4} & -r^2 \\ -r^2 & 1 + \frac{r^4}{4} \end{pmatrix}.$$

By (4.15), we have

$$\begin{pmatrix} \widetilde{\mathbf{c}}_{13} & \widetilde{\mathbf{c}}_{14} \\ \widetilde{\mathbf{c}}_{23} & \widetilde{\mathbf{c}}_{24} \\ \widetilde{\mathbf{c}}_{33} & \widetilde{\mathbf{c}}_{34} \\ \widetilde{\mathbf{c}}_{43} & \widetilde{\mathbf{c}}_{44} \\ \widetilde{\mathbf{c}}_{53} & \widetilde{\mathbf{c}}_{54} \\ \widetilde{\mathbf{c}}_{63} & \widetilde{\mathbf{c}}_{64} \end{pmatrix} = \frac{1}{\varsigma}\begin{pmatrix} \frac{-\lambda^{-1}zr^2}{2} & \lambda^{-1}z \\ \lambda^{-1}z & -\frac{\lambda^{-1}zr^2}{2} \\ 1 + \frac{r^4}{4} & -r^2 \\ -r^2 & 1 + \frac{r^4}{4} \\ \lambda\bar{z} & -\frac{\lambda\bar{z}r^2}{2} \\ -\frac{\lambda\bar{z}r^2}{2} & \lambda\bar{z} \end{pmatrix}.$$

Substituting these data into (4.27), one derives (4.21). The rest are straightforward computations, which we will leave to interested readers.

Acknowledgments

The author was supported by the NSFC Project No. 11571255.

Bibliography

[1] Brander, D., Wang, P. *On the Björling problem for Willmore surfaces*, to appear in J. Diff. Geom.

[2] Bryant, R. *Conformal and minimal immersions of compact surfaces into the 4-sphere*, J. Diff. Geom. 17(1982), 455-473.

[3] Bryant, R. *A duality theorem for Willmore surfaces*, J. Diff. Geom. 20 (1984), 23-53.

[4] Burstall, F.E., Guest, M.A., *Harmonic two-spheres in compact symmetric spaces, revisited*, Math. Ann. 309 (1997), 541-572.

[5] Burstall, F., Pedit, F., Pinkall, U. *Schwarzian derivatives and flows of surfaces*, Contemporary Mathematics 308, 39-61, Providence, RI: Amer. Math. Soc., 2002.

[6] Burstall, F., Quintino, A., *Dressing transformations of constrained Willmore surfaces*, Comm. Anal. Geom. 22 (2014), no. 3, 469-518.

[7] Calabi, E. *Minimal immersions of surfaces in Euclidean spheres*, J. Diff. Geom. 1(1967), 111-125.

[8] Dorfmeister, J., Pedit, F., Wu, H., *Weierstrass type representation of harmonic maps into symmetric spaces*, Comm. Anal. Geom. 6 (1998), 633-668.

[9] Dorfmeister, J., Wang, P., *Willmore surfaces in spheres via loop groups I: generic cases and some examples*, arXiv:1301.2756v4.

[10] Dorfmeister, J., Wang, P., *Harmonic maps of finite uniton type into non-compact inner symmetric spaces*, arXiv:1305.2514v2.

[11] Ejiri, N., *Willmore surfaces with a duality in $S^n(1)$*, Proc. London Math.Soc. (3), 57(2) (1988), 383-416.

[12] Guest, M.A. *An update on Harmonic maps of finite uniton number, via the Zero Curvature Equation*, Integrable Systems, Topology, and Physics: A Conference on Integrable Systems in Differential Geometry

(Contemp. Math., Vol. 309, M. Guest et al., eds.), Amer. Math. Soc., Providence, R. I. (2002), 85-113.

[13] Hélein, F. *Willmore immersions and loop groups,* J. Differ. Geom., 50, 1998, 331-385.

[14] Hélein, F. *A Weierstrass representation for Willmore surfaces,* Harmonic morphisms, harmonic maps, and related topics (Brest, 1997), vol. 413 of Res. Notes Math., Chapman Hall/CRC, Boca Raton, FL, 2000, pp. 287-302.

[15] Ma, X. *Adjoint transforms of Willmore surfaces in* S^n, Manuscripta Math., 120 (2006), 163-179.

[16] Ma, X. *Willmore surfaces in* S^n*: transforms and vanishing theorems,* Ph.D. dissertation, TU Berlin, 2005.

[17] Ma, X., Wang, C. P., Wang, P. *Classification of Willmore 2-spheres in the 5-dimensional sphere*, to appear in J. Diff. Geom.

[18] Ma, X., Wang, P. *On Willmore two-spheres in* S^6, in preparation.

[19] Montiel, S. *Willmore two spheres in the four-sphere,* Trans. Amer.Math. Soc. 2000, 352(10), 4469-4486.

[20] Musso, E. *Willmore surfaces in the four-sphere,* Ann. Global Anal. Geom. Vol 8, No.1(1990), 21-41.

[21] Pressley, A.N., Segal, G.B., *Loop Groups,* Oxford University Press, 1986.

[22] Uhlenbeck, K. *Harmonic maps into Lie groups (classical solutions of the chiral model),* J. Diff. Geom. 30 (1989), 1-50.

[23] Wang, C.P. *Moebius geometry of submanifolds in* S^n, Manuscripta Math., 96 (1998), No.4, 517-534.

[24] Wang, P., *Willmore surfaces in spheres via loop groups II: a coarse classification of Willmore two-spheres via potentials*, arXiv:1412.6737.

[25] Wang, P., *Willmore surfaces in spheres via loop groups IV: on totally isotropic Willmore two-spheres in* S^6, arXiv:1412.8135.

[26] Wu, H.Y. *A simple way for determining the normalized potentials for harmonic maps,* Ann. Global Anal. Geom. 17 (1999), 189-199.

[27] Xia, Q.L., Shen, Y.B. *Weierstrass type representation of Willmore surfaces in* S^n. Acta Math. Sin. (Engl. Ser.), 20 (2004), No. 6, 1029-1046.

Chapter 5

Towards a Constrained Willmore Conjecture

Lynn Heller, Franz Pedit

CONTENTS

Abstract

We give an overview of the constrained Willmore problem and address some conjectures arising from partial results and numerical experiments. Ramifications of these conjectures would lead to a deeper understanding of the Willmore functional over conformal immersions from compact surfaces.

5.1 Introduction

When bending an elastic membrane its bending energy per area can be shown to be locally proportional to the square of its (mean) curvature. Very likely this has been known to Hooke as a manifestation of his *stress proportional to strain law*. Mathematically we model the membrane by an immersed surface $f\colon M \to \mathbb{R}^3$. The bending energy density is then given by $H^2 dA$ with H the mean curvature (the arithmetic mean of the principal curvatures) and

dA the area form both calculated with respect to the induced Riemannian metric $|df|^2$ on M. Thus, the total *bending energy* is

$$\mathcal{W}(f) = \int_M H^2 dA$$

which is also known as the *Willmore energy* of the immersion f. Indeed, in the mid 1960s Willmore [71] proposed to study the minima and also the stationary points, *Willmore surfaces*, of the functional $\mathcal{W}$ over immersions f of compact surfaces M. He sought for an extrinsic analogue to the notion of tight immersions which are the minima of the total absolute Gaussian curvature $\int_M |K| dA$. Willmore showed that $\mathcal{W}(f) \geq 4\pi$ with equality attained at a round sphere of any radius since $\mathcal{W}(f)$ is invariant under scaling. Furthermore, Willmore verified that among tori of revolution in $\mathbb{R}^3$ the Clifford torus (and its scaled versions), whose profile circle of radius 1 has its center at distance $\sqrt{2}$ from the rotation axis, is the minimum with energy $2\pi^2$. He then posed the question, later to become the *Willmore conjecture*, whether the infimum of $\mathcal{W}$ over immersed surfaces of genus one is $2\pi^2$ and whether it is attained at the Clifford torus.

Seemingly unknown to Willmore, in the 1920s Blaschke and his school [4] studied surface theory in Möbius geometry. Among other things they showed that one can associate to a surface $f\colon M \to \mathbb{R}^3$ in a Möbius invariant way a *conformal Gauss map* by attaching to any point $f(p)$ the sphere tangent to the surface of radius $1/H(p)$. Working in the light cone model of the conformal compactification $\mathbb{R}^3 \cup \{\infty\} \cong S^3$, the space of oriented 2-spheres is parameterized by the Lorentzian 4-sphere S^4_1, a pseudo-Riemannian rank 1 symmetric space of constant curvature one. The Möbius group of 3-space is presented by the causality preserving isometries $\mathbf{O}^+(4,1)$ of S^4_1. The Möbius invariant induced area form of the conformal Gauss map $\gamma_f\colon M \to S^4_1$ can be written as $(H^2 - K)dA$. This motivated Blaschke and his school to study "conformally area minimizing" surfaces $f\colon M \to \mathbb{R}^3$, that is the critical points of the functional

$$\tilde{\mathcal{W}}(f) = \int_M (H^2 - K) dA$$

We note that on compact surfaces of genus g the functionals $\mathcal{W}$ and $\tilde{\mathcal{W}}$ differ by the Euler number $\int_M K dA = 4\pi(1-g)$ and thus have the same critical points: Willmore surfaces are Blaschke's conformal minimal surfaces and therefore invariant under the full group of Möbius transformations of 3-space.

Some of the, mostly local, results obtained by Blaschke and his collaborators have eventually been rediscovered or reproven:

- The Euler-Lagrange equation of $\mathcal{W}$ is given by $\triangle H + 2H(H^2 - K) = 0$. Thus Möbius transforms of minimal surfaces are examples of Willmore surfaces;
- A surface $f\colon M \to \mathbb{R}^3$ is Willmore if and only if its conformal Gauss map $\gamma_f\colon M \to S^4_1$ is (branched) minimal;

- Isothermic Willmore surfaces are (locally) Möbius congruent to a minimal surface in one of the space forms [70].

The integrant $H^2 - K = |\mathring{S}|^2$ is just the squared length of the tracefree 2nd fundamental form of the immersion $f\colon M \to \mathbb{R}^3$, which is known to scale inverse proportionally to the induced area under Möbius transformations (in fact, under any conformal changes of the ambient metric). Hence $\tilde{\mathcal{W}} = \int_M |\mathring{S}|^2 dA$ is defined for any conformal target manifold. In particular, ignoring the Euler number, we have $\mathcal{W} = \int_M (H^2 \pm 1) dA$ for immersions into the sphere S^3 and hyperbolic space H^3.

In the early 1980s Bryant [11] classified all Willmore 2-spheres $f\colon S^2 \to \mathbb{R}^3$ as Möbius inversions of minimal surfaces with planar ends in $\mathbb{R}^3$. The Willmore energy $\mathcal{W} = 4\pi k$ is quantized by the number of ends and, except for $k = 2, 3, 5, 7$, all values are realized. For $k = 1$ one obtains the absolute minimum of $\mathcal{W}$, a round 2-sphere or, after an inversion, a plane. For higher values of $\mathcal{W}$ there are smooth families of Willmore spheres.

The classification of Willmore tori and a resolution of the Willmore conjecture turned out to be more involved. The first lower bound for $\mathcal{W}$ over immersions f of a Riemann surface M into the n-sphere S^n was given in early 1980 in a paper by Li and Yau [51]. Due to Möbius invariance of the Willmore energy one has

$$\mathcal{W}(f) = \int_M (H^2 + 1) dA \geq \sup_T \mathcal{A}(T \circ f)$$

Here T ranges over all Möbius transformations of S^n and $\mathcal{A}$ denotes the area functional. Therefore, one can estimate the minimal Willmore energy for a fixed conformal class from below by the conformal area $\inf_f \sup_T \mathcal{A}(T \circ f)$, where f ranges over all conformal immersions of M into S^n. Applying a first eigenvalue estimate for the Laplacian, Li and Yau could show that for a torus $M = \mathbb{R}^2/\mathbb{Z} \oplus \mathbb{Z}\tau$ whose conformal structure lies in the domain $|\tau| \geq 1$, $\mathrm{Im}\tau \leq 1$, $0 \leq \mathrm{Re}\tau \leq 1/2$ the conformal volume is bounded from below by $2\pi^2$. In their study of minimal immersions given by first eigenfunctions of the Laplacian, Montiel and Ros [57] (and independently Bryant [12]) enlarged this domain slightly. They also provided examples of immersions of tori of rectangular conformal types whose conformal areas were below $2\pi^2$, thus erasing hopes that this approach would resolve the Willmore conjecture. Li and Yau also showed the useful estimate that if an immersion has a multiple point of order k then $\mathcal{W}(f) \geq 4\pi k$.

The existence of a minimizer for $\mathcal{W}$ over immersions of tori was first proven by Simon [69] in the early 1990s, and then in early 2000 by Bauer and Kuwert [3] for any genera and in any codimension. But whether the minimizer in genus one had to satisfy $\mathcal{W} \geq 2\pi^2$ remained open. Besides the cases covered in [51] the Willmore conjecture had been verified for channel tori (envelopes of a circle worth of round 2-spheres) [35] and tori invariant under an antipodal symmetry [65]. Numerical evidence for the validity of the conjecture was obtained for Hopf Willmore tori [61], which were also the first examples of Willmore surfaces not Möbius congruent to minimal surfaces, and for S^1-equivariant Willmore tori [22].

In the early 1990s the emerging integrable system techniques used to study harmonic maps from tori into symmetric spaces [33], [62], [23], [13] began to be applied to the conformal Gauss maps of Willmore surfaces. Even though a deeper appreciation of the complexity of the problem was obtained, for instance in the unpublished work by Schmidt [67] in early 2000, and new lower bounds in terms of the integrable systems structure were found, notably the quaternionic Plücker formula [21], this approach has so far failed to provide a resolution of the Willmore conjecture. It was not until early 2012 when Marques and Neves [55] gave a proof of the Willmore conjecture using very different techniques based on a refinement of the Almgren and Pitts Min-Max approach.

Knowing the existence of minimizers for any genus and the minimal value for $\mathcal{W}$ over immersions $f\colon M \to \mathbb{R}^3$ of tori, we can ask what to expect in higher genus. So far the only known examples of higher genus $g \geq 2$ Willmore surfaces in $\mathbb{R}^3$ are stereographic projections of compact minimal surfaces in the 3-sphere. The most prominent examples are Lawson's minimal surfaces $\xi_{g,1}$ of any genus g with dihedral symmetries [49], which naturally lend themselves as candidates for the minimizers [41]. Their Willmore energies $\mathcal{W}$ are increasing with the genus, starting from the Clifford torus at $\mathcal{W} = 2\pi^2$ and limiting to 8π as the genus tends to infinity. There is some experimental evidence [34], using bending energy decreasing flows, that Lawson's surfaces are indeed the minima for $\mathcal{W}$ on higher genus surfaces.

In addition to the existence results of minimizers in arbitrary codimension [69, 3] there has been renewed interest in the study of Willmore surfaces in higher codimension. In the late 1980s Ejiri [20] gave a description of Willmore 2-spheres admitting a dual Willmore 2-sphere in any codimension in terms of holomorphic lifts to a suitable twistor space. In codimension two this was made explicit in a paper by Montiel [58], who showed that every Willmore 2-sphere in codimension two is either the inversion of a minimal surface with planar ends in $\mathbb{R}^4$ or the projection of a rational curve under the Penrose twister fibration $\mathbb{CP}^3 \to S^4$. The Willmore energy is quantized by $\mathcal{W} = 4\pi k$ and all $k \in \mathbb{Z}$ occur. This result holds verbatim for Willmore tori in codimension two, provided the normal bundle has non-zero degree [50]. Applying loop group techniques Dorfmeister and Wang [19] recover some of those results in their language and also provide a general setting for studying Willmore surfaces in the framework introduced in [18]. A more geometric approach, in the spirit of Calabi's characterization [15] of minimal 2-spheres, is given by Ma and his collaborators [52], [53], [54]. This approach studies generalized Darboux transforms of Willmore surfaces and aims to construct a sequence of transforms which terminate, at least for Willmore 2-spheres, in a minimal surface with planar ends or a twistor projection of a rational curve. Even though the existence of minimizers is known for any genus and in any codimension, it is still an open question whether the Willmore conjecture holds in higher codimension. It is conceivable, though hard to imagine, that a minimizing torus in $\mathbb{R}^4$ has Willmore energy below $2\pi^2$.

A refinement of the Willmore problem arises if we fix the conformal structure of the domain M and look for minimizers (and more generally critical points) of the Willmore functional $\mathcal{W}$ under variations preserving the conformal type of M. The critical points are called (conformally) *constrained Willmore surfaces*. From the perspective of the integrable systems approach working in a fixed conformal class is more natural, if not required, since the conformal type is one of the integrals of motion. There is also some hope that by utilizing Riemann surface theory - especially the description of conformally immersed surfaces in 3- and 4-space by a Dirac equation with potential [7] - the 4th order non-linear elliptic analysis of the Euler-Lagrange equation can, to some extent, be replaced by the 1st order linear analysis of the Dirac equation. Such ideas may be found in the unpublished work of Schmidt [67], [68] but have yet to be made precise.

The conformal constraint introduces, as a Lagrange multiplier, a quadratic holomorphic differential $Q \in H^0(K^2)$ into the Euler-Lagrange equation

$$\triangle H + 2H(H^2 - K) = < Q, \mathring{S} >$$

where $\mathring{S}$ denotes the tracefree 2nd fundamental form. Choosing $Q = 0$ shows that every Willmore surface is a constrained Willmore surface. For a constant mean curvature surface we choose Q to be its holomorphic Hopf differential to verify that those surfaces (and their Möbius transforms) are constrained Willmore. Since there are no holomorphic differentials on a genus zero Riemann surface, constrained Willmore spheres are Willmore spheres. There is a subtlety when deriving the constrained Euler-Lagrange equation: the subspace of conformal immersions of a Riemann surface inside the smooth manifold of all immersions is singular exactly at isothermic immersions. Since many of the known examples are isothermic (for instance, constant mean curvature tori) the validity of the equation at those surfaces required a deeper analysis of second variations [66].

Existence of constrained minimizers in any conformal class has been proven for any genus and any codimension [46], [42], [64] provided that the given conformal class can be realized by an immersion of Willmore energy below 8π. This condition is used to avoid that the minimizer acquires branch points (a branch point - as a limit of a double point - would imply that $W \geq 8\pi$ by the estimate of Li and Yau). In particular, the constrained minimizers with Willmore energy below 8π are embedded tori. In addition, the Willmore energy $\mathcal{W}$ varies continuously over the conformal types of those minimizers [46]. Numerical experiments with bending energy decreasing flows [30], [17] suggest that constrained minimizers of genus one, whose conformal structures are sufficiently far from rectangular structures, will have $W \geq 8\pi$. If one allows codimension at least two, then every conformal type [48] can be realized by a conformal immersion with Willmore energy $\mathcal{W} \leq 8\pi$. There is some evidence [68] that in codimension one the upper bound is 12π.

On the other hand, independent of the proof of the Willmore conjecture [55], Ndiaye and Schätzle [59] show that for rectangular conformal types $\mathbb{R}^2/\mathbb{Z} \oplus ib\mathbb{Z}$ sufficiently near to the square structure the minimizers in $\mathbb{R}^3$ are the (stereographic projection of) homogeneous flat tori $S^1 \times S^1(1/b)$ in 3-spheres. They also generalize this result to arbitrary codimension [60].

After this somewhat broad stroked and kaleidoscopical overview of the subject, we will discuss some of the emerging conjectures concerning constrained Willmore surfaces together with supporting evidence and partial results.

Acknowledgments: part of this research was supported by the German Science Foundation DFG. The first author is indebted to the Baden-Württemberg Foundation for supporting her research through the "Elite Program for Postdocs". The second author was additionally supported by a University of Massachusetts professional development grant. All images were produced by Dr. Nicholas Schmitt using his Xlab software suite.

5.2 The constrained Willmore conjecture

The classical Willmore conjecture asserts that among immersions of tori in 3-space the Clifford torus is the absolute minimum with Willmore energy $2\pi^2$. The Clifford torus is also the unique embedded minimal torus in the 3-sphere, a result proven recently by Brendle [10] and conjectured in the 1970s by Lawson [49]. Since surfaces of constant mean curvature are constrained Willmore, it might be reasonable to guess that the constrained minimizers in a given conformal class could be found among constant mean curvature tori. We already saw that the existence of constrained minimizers [46] is guaranteed only for surfaces with $\mathcal{W} < 8\pi$ (in which case they have to be embedded). Hence one should first consider embedded constant mean curvature tori as candidates. By a recent result of Andrews and Li [1] based on ideas in [10], such surfaces have to be rotational tori in the 3-sphere. This result had been conjectured by Pinkall and Sterling [62] as a generalization of Lawson's conjecture. Embeddedness restricts those surfaces to be the products of circles, the homogeneous tori $S^1 \times S^1(b)$, for any $b \geq 1$, and the unduloidal k-lobed Delaunay tori [39] in a 3-sphere for $k \geq 2$ (see Figure 5.2). The conformal types $\mathbb{R}^2/\mathbb{Z} \oplus \tau\mathbb{Z}$ of those tori sweep out all rectangular conformal types $\tau = ib$ for $b \in [1, \infty)$ starting at the square structure. The k-lobed Delaunay tori branch off from homogeneous tori at the conformal type $\tau = i\sqrt{k^2 - 1}$ and limit to a bouquet of k spheres of degenerate conformal type. The Willmore energy $\mathcal{W}(S^1 \times S^1(b)) = \pi^2(b + 1/b)$ of homogenous tori is unbounded and hence they cannot minimize $\mathcal{W}$ for all rectangular conformal types. In fact, Kuwert and Lorenz [43] calculate the Jacobi operator for $\mathcal{W}$ along homogeneous tori and show that negative

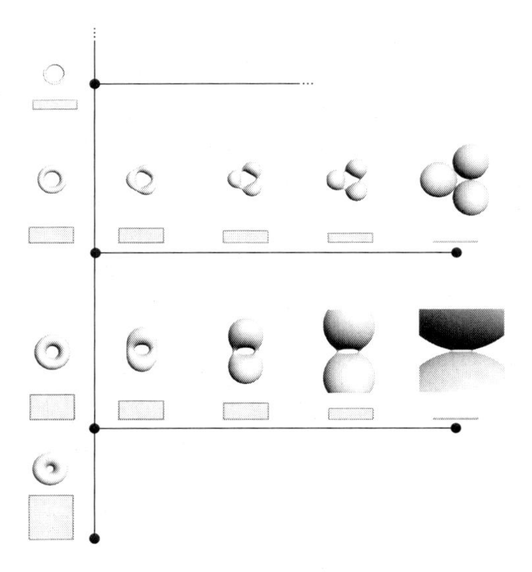

FIGURE 5.1: The vertical stalk represents the family of homogenous tori, starting with the Clifford torus at the bottom. Along this stalk are bifurcation points at which the embedded Delaunay tori appear along the horizontal lines. The rectangles indicate the conformal types.

eigenvalues appear precisely at $b = \sqrt{k^2 - 1}$ for $k \geq 2$, where the Delaunay tori branch off. Combining arguments in [39] and [40] the Willmore energy along the family of homogeneous tori $S^1 \times S^1(b)$ for $1 \leq b \leq \sqrt{k^2 - 1}$ and continuing along the unduloidal k-lobed Delaunay tori is monotonically increasing between $2\pi^2 \leq \mathcal{W} < 4\pi k$ (the latter being the Willmore energy of the bouquet of k-spheres).

Thus, considering the case $k = 2$, we have an embedded constant mean curvature torus with $2\pi^2 \leq \mathcal{W} < 8\pi$ for every rectangular conformal type $\mathbb{R}^2/\mathbb{Z} \oplus ib\mathbb{Z}$ which minimizes $\mathcal{W}$ among constant mean curvature tori in those conformal classes. Taking into account the results of Ndiaye and Schätzle [59], [60], who showed that for rectangular conformal types near the square structure the homogeneous tori $S^1 \times S^1(b)$ minimize, we are led to a constrained Willmore conjecture for rectangular conformal types:

Conjecture 1 (2-lobe conjecture). *The constrained minimizers of the Willmore energy for tori in $\mathbb{R}^3$ of rectangular conformal types $\mathbb{R}^2/\mathbb{Z} \oplus ib\mathbb{Z}$ are (stereographic projections of) the homogeneous tori $S^1 \times S^1(b)$ in the 3-sphere for $1 \leq b \leq \sqrt{3}$ and the 2-lobed Delaunay tori in a 3-sphere for $b > \sqrt{3}$ limiting to a twice covered equatorial 2-sphere as $b \to \infty$ (see Figure 5.2).*

For constrained minimizers in non-rectangular conformal classes there is much less guidance due to a lack of examples. Constrained minimizers, whose non-rectangular conformal types lie in a sufficiently small open neighborhood of the square structure will, by continuity, still have Willmore energy below 8π and thus have to be embedded. Therefore, they cannot have constant mean curvature by our previous discussion. The first examples of constrained Willmore tori *not* Möbius congruent to constant mean curvature tori were constructed by Heller [24], [25], who described all S^1 equivariant constrained Willmore tori in the 3-sphere where the orbits of the action are (m, n) torus knots. For instance, $(1, 1)$-equivariance yields constrained Hopf Willmore tori which arise as preimages under the Hopf projection $S^3 \to S^2$ of length and enclosed area constrained elastica on the round 2-sphere of curvature 4. Their conformal types $\tau = a + ib$ are determined by the enclosed areas a and lengths b of the elastica and all conformal types occur. In a recent preprint Heller and Ndiaye [29] generalize the results in [59] and characterize the minima of the constrained Willmore problem in a certain neighborhood of rectangular conformal classes:

Theorem. *For every $b \sim 1$ and $b \neq 1$ there exists a sufficiently small $a(b) > 0$ such that for every $a \in [0, a(b))$ the $(1, 2)$ equivariant constrained Willmore torus $f_{a,b}$ of intrinsic period 1 (see Figure 5.3) and conformal type $\tau = a + ib$ is a constrained Willmore minimizer. Moreover, for fixed $b \neq 1$ and $b \sim 1$ the energy profile $a \mapsto \mathcal{W}(f_{a,b})$ is concave and varies real analytically over $[0, a(b))$.*

It is expected that this result can be extended to a neighborhood of all rectangular conformal types. In fact, laying a grid over the space of conformal structures and using an implementation of the conformal Willmore flow [17], we saw evidence that the $(1, 2)$-equivariant constrained Willmore tori minimize [30] in all conformal classes (see Figure 5.3). These experiments also suggest that for conformal structures sufficiently far from rectangular types these tori have Willmore energy $\mathcal{W} \geq 8\pi$ and in general will not be embedded (see Figure 5.2). We emphasize though that for such conformal structures there is, as of yet, no existence proof guaranteeing immersed minimizers. Based on these observations we are led to

Conjecture 2 (Constrained Willmore Conjecture)**.** *The minimizers of the Willmore functional $\mathcal{W}$ over tori in a prescribed conformal class are given by the $(1, 2)$-equivariant constrained Willmore tori (of intrinsic period 1).*

In contrast to the results in [60] for rectangular conformal classes near the square structure, the above theorem cannot be generalized to higher codimension: it relies on the stability computations in [43] for codimension 1. We do expect to find other minimizers in higher codimension: for instance the minimal torus of hexagonal conformal type in the 5-sphere [57] of Willmore energy $\mathcal{W} = 4\pi^2/\sqrt{3} < 8\pi$. In fact, as already mentioned, in codimension at least 2 every conformal type [48] can be realized by a conformal immersion with Willmore energy $\mathcal{W} \leq 8\pi$.

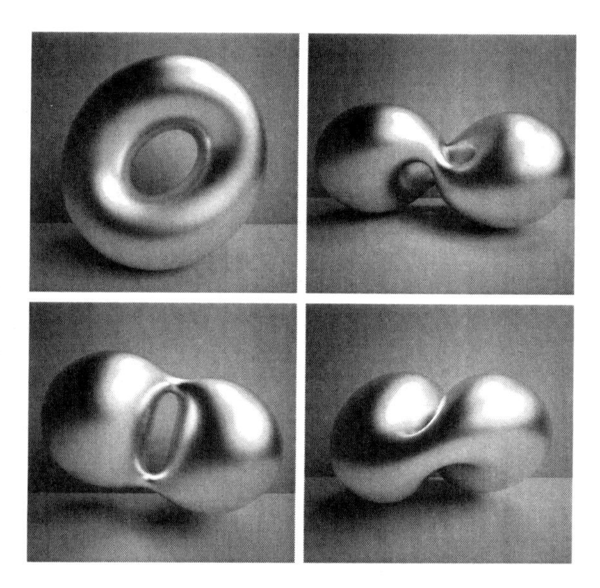

FIGURE 5.2: Experimentally found minimizers of the constrained Willmore problem [30]. They resemble $(1,2)$-equivariant constrained Willmore tori with intrinsic period 1 - compare to Figure 5.3. The conformal structures $\tau = a + ib$ of the shown tori are as follows: upper row has $b < \sqrt{3}$, lower row has $b > \sqrt{3}$; left column has $a \sim 0$, right column has $a >> 0$. The torus on the bottom right is not embedded.

5.3 The constrained Willmore Lawson conjecture

In the above discussions we have seen that having a rich enough class of examples at one's disposal helps guide the theory. Understanding $(1,2)$-equivariant tori and comparing them to the experimentally found minimizers gave some credence to the constrained Willmore conjecture. Unfortunately, so far the only known examples of compact constrained Willmore surfaces are equivariant, or surfaces Möbius congruent to constant mean curvature surfaces. To obtain a more detailed understanding of constrained Willmore surfaces a wider class of examples seems desirable. Since the conformal Gauss map $\gamma_f \colon M \to S^4_1$ of a Willmore surface $f \colon M \to \mathbb{R}^3$ is a harmonic (in fact, branched minimal) map into the space of round 2-spheres, one can apply the integrable systems theory for harmonic maps into symmetric target spaces to construct examples of Willmore tori.

FIGURE 5.3: $(1,2)$-equivariant constrained Willmore tori with intrinsic period 1. The tori lie in a 2 parameter family of constrained Willmore tori deforming the Clifford torus and minimize the Willmore functional in conformal classes near the square structure.

The associated family of the harmonic map γ_f into the Lorentz 4-sphere S^4_1 gives rise to a $\mathbb{C}^\times$ family of flat $\mathbf{sl}(4,\mathbb{C}) \cong \mathbf{so}(5,\mathbb{C})$ connections [18], [19], [23]

$$\nabla^\lambda = \nabla + \lambda^{-1}\Phi + \lambda\bar{\Phi}$$

Here $\Phi \in \Omega^{1,0} \otimes \mathbf{sl}(4,\mathbb{C})$ is given by the $(1,0)$ part $\Phi = d\gamma_f^{1,0}$ of the derivative of γ_f and ∇ is the pullback of the Levi-Civita connection on S^4_1. There is a reality condition, namely for $\lambda \in S^1 \subset \mathbb{C}^\times$ the ∇^λ have to be $\mathbf{sp}(1,1)$ connections. In this reformulation of the Euler-Lagrange equation for Willmore surfaces one can augment the loop ∇^λ by a quadratic holomorphic differential Q and obtain a verbatim characterization of constrained Willmore surfaces as a family of flat connections [14], [6]. The conformal Gauss map γ_f then is a constrained branched minimal map. In the case when $M = T^2$ is a 2-torus the flat connection ∇^λ has four distinct parallel line subbundles except at isolated values of $\lambda \in \mathbb{C}^\times$ where pairs of the line bundles coalesce. Bohle [6] has shown that there can only be finitely many such points and that the line bundles have limits as $\lambda \to 0, \infty$. Thus, there exists a compact Riemann surface $\lambda\colon \Sigma \to \mathbb{P}^1$, quadruply covering $\mathbb{P}^1$, parametrizing the parallel line bundles. The Riemann surface Σ is called the *spectral curve* and its genus g_s the *spectral genus* of the constrained Willmore torus. Now a point $L \in \Sigma$ presents a parallel line bundle for $\nabla^{\lambda(L)}$ and the family of connections ∇^λ is holomorphic in λ. Therefore, for fixed $p \in T^2$, the fibers L_p, as L varies over Σ, fit to a holomorphic line bundle $L(p) \to \Sigma$. The complete integrability of constrained Willmore tori is encoded in the statement that the map

$$T^2 \to \mathrm{Jac}(\Sigma)\colon p \to L(p)$$

is a group homomorphism - linear flow - of prescribed direction from the 2-torus T^2 into the (real) Jacobi torus $\mathrm{Jac}(\Sigma)$ of holomorphic line bundles over

Σ. Since a group homomorphism of a given slope is determined by its value at one point $L(p_0)$, we can associate to a Willmore torus $f\colon T^2 \to \mathbb{R}^3$ its *spectral data*: the spectral curve Σ and an initial condition $L(p_0) \in \mathrm{Jac}(\Sigma)$ for the linear flow. It has been shown in [7], [8] that one can reconstruct the constrained Willmore torus up to Möbius equivalence from its spectral data. Notice that once Σ is chosen there is a $g_s = \dim \mathrm{Jac}(\Sigma)$ dimensional freedom of choosing the initial condition $L(p_0)$ of the flow. This accounts for *isospectral* deformations of a constrained Willmore torus: two of those dimensions conformally reparametrize the surface, but the $g_s - 2$ dimensions transverse to the linear flow $T^2 \to \mathrm{Jac}(\Sigma)$ yield non Möbius congruent constrained Willmore tori (necessarily of the same conformal type and Willmore energy).

What has been described here is a geometric manifestation of the finite gap theory of the Novikov-Veselov hierarchy: the isospectral deformations account for the higher flows of the hierarchy. In the special case of constant mean curvature tori the spectral curve $\Sigma \to \mathbb{P}^1$ is hyperelliptic. This case, corresponding to the sinh (or cosh) Gordon hierarchy, has been studied in great detail [62], [33], [5] and many examples of higher spectral genus $g_s \geq 2$ constant mean curvature tori are known (for instance, the Wente and Dobriner tori).

The spectral genus g_s gives some information about the complexity of the corresponding constrained Willmore torus [24], [25]:

- $g_s = 0$ characterizes the homogeneous tori $S^1 \times S^1(b)$ for $b \in [1, \infty)$;

- $g_s = 1$ characterizes equivariant constant mean curvature tori constructed in [39];

- $g_s = 2$ characterizes surfaces Möbius congruent to constant mean curvature tori of Wente type in any of the three space forms, and equivariant constrained Willmore tori in the associated family of constrained Hopf Willmore cylinders. In particular, the conjectured constrained minimizers in a given conformal class, the $(1, 2)$-equivariant constrained Willmore tori, are found here.

In general, equivariant constrained Willmore tori of orbit type (m, n) have spectral genus $g_s \leq 3$ and there is a complete classification [24], [25] of these surfaces for the case $g_s \leq 2$, including explicit parametrizations.

The lowest spectral genus where one can find constrained Willmore tori, which are neither equivariant nor Möbius congruent to constant mean curvature tori, has to be at least $g_s \geq 3$, but its value is unknown. The reconstruction of a constrained Willmore torus from spectral data is in principle possible [7], [8] even though the technical details have not been worked out. To construct such new examples of constrained Willmore tori and to understand their structure would significantly increase our confidence in

Conjecture 3 (Constrained Willmore Lawson conjecture). *An embedded constrained Willmore torus is equivariant.*

We know that the conjecture holds for an embedded constrained Willmore torus which is Möbius congruent to a constant mean curvature torus by the results of Brendle [10] and Li and Andrews [1]. An affirmative answer to this conjecture has significant ramifications:

- *The classical Willmore conjecture*: since the (unconstrained) minimizer of the Willmore functional has to be embedded, it would have to be equivariant. But the Willmore conjecture holds (or at least can be verified) for equivariant Willmore tori.
- *The 2-lobe conjecture*: since the constrained minimizers in rectangular conformal classes have to be embedded, it would suffice to study the equivariant case to resolve the constrained Willmore conjecture for rectangular conformal types.

5.4 The stability conjecture

We are now entering into much less charted terrain. There is some evidence that Willmore energy decreasing flows on immersed 2-spheres limit to the round sphere. This has been shown using long time existence of the Willmore gradient flow on 2-spheres with $\mathcal{W} \leq 8\pi$ in [45], [44] and is suggested by experiments in [17] using a conformal Willmore flow. Therefore one is led to believe that the only stable Willmore spheres are round 2-spheres. In what follows stability always refers to the positive (semi) definiteness $\delta^2\mathcal{W} \geq 0$ of the Jacobi operator for the Willmore functional.

Although there is less supporting theoretical evidence for surfaces of genus one, we formulate

Conjecture 4 (Stability conjecture). *The only stable Willmore torus is the Clifford torus.*

As we have explained in the previous section, every Willmore torus $f\colon M \to \mathbb{R}^3$ has some spectral genus g_s and is contained in the $g_s - 2$ dimensional isospectral family of non Möbius congruent Willmore tori. If V denotes the kernel (modulo Möbius tranformations) of the Jacobi operator $\delta^2\mathcal{W}_f$, then we have

$$\dim V \geq g_s - 2$$

Therefore, a *strictly stable* Willmore torus $f\colon M \to \mathbb{R}^3$, that is f is stable and V is trivial, has spectral genus $g_s \leq 2$. All Willmore tori of spectral genus $g_s \leq 2$ are known [25] to be either Möbius congruent to a minimal torus of Wente type in one of the space forms, or associated to the Hopf Willmore surfaces found by Pinkall [61]. These surfaces could in principle be checked case by case to affirm the stability conjecture for strictly stable Willmore tori.

There is further support for the conjecture involving the multiplicity μ_1 (modulo Möbius transformations) of the first eigenvalue λ_1 of the Jacobi operator $\delta^2\mathcal{W}_f$. Since stability implies $\lambda_1 \geq 0$, we have either $\dim V = 0$ (for $\lambda_1 > 0$) or $\dim V = \mu_1$ (for $\lambda_1 = 0$). Therefore a stable Willmore torus has spectral genus at most $g_s \leq \mu_1 + 2$. For example, if one could show (as is the case for the Laplacian) that $\mu_1 = 1$ then the stability conjecture could be verified by checking only those Willmore tori of spectral genus $g_s = 3$.

Among Hopf Willmore tori [61] the stability conjecture follows from the fact that the analogous conjecture holds for elastic curves, the critical points of the total squared curvature: stable elastica on the 2-sphere have to be great circles [47]. Thus, the only stable Hopf Willmore torus [61] is the Clifford torus.

The following weaker version of the stability conjecture, which is of interest in its own right, would already provide an alternative proof of the Willmore conjecture:

Conjecture 5 (Weak stability conjecture)**.** *A stable Willmore torus has to be isothermic.*

Assuming the conjecture to hold, Thomsen's result [70] implies that an isothermic Willmore torus is (locally) Möbius congruent to a minimal surface in one of the space forms. If the space form is S^3 then, due to the fact that a Willmore minimizer is embedded, we obtain an embedded minimal torus. In this case Brendle's resolution [10] of the Lawson conjecture [49] implies that the minimizer has to be the Clifford torus. The other possibility is that the Willmore torus is an inversion of a minimal torus with planar ends in $\mathbb{R}^3$. But those surfaces are never embedded [9]. Finally, the Willmore torus could be one of the minimal tori coming from doubly periodic solutions of the cosh Gordon equation: these tori consist of two minimal pieces contained in the upper respectively lower hyperbolic half spaces H^3 smoothly connecting across the boundary at infinity, along which the torus has umbilic curves. The known examples [2] are not embedded but it is unknown whether all such tori arising from cosh Gordon solutions have self intersections.

5.5 Higher genus outlook

Examples of compact constrained Willmore surfaces of genus $g \geq 2$ are scarce: the $\mathbb{Z}_{k+1} \times \mathbb{Z}_{l+1}$ symmetric compact minimal surfaces $\xi_{k,l}$ in the 3-sphere of genus $g = kl$ found by Lawson [49], the minimal surfaces in the 3-sphere with Platonic symmetries constructed by Karcher, Pinkall and Sterling [38], and the constant mean curvature surfaces of any genus constructed via gluing methods by Kapouleas [37], [36]. Among Lawson's minimal surfaces Kusner [41] verified that the family of genus g surfaces $\xi_{g,1}$ have the lowest

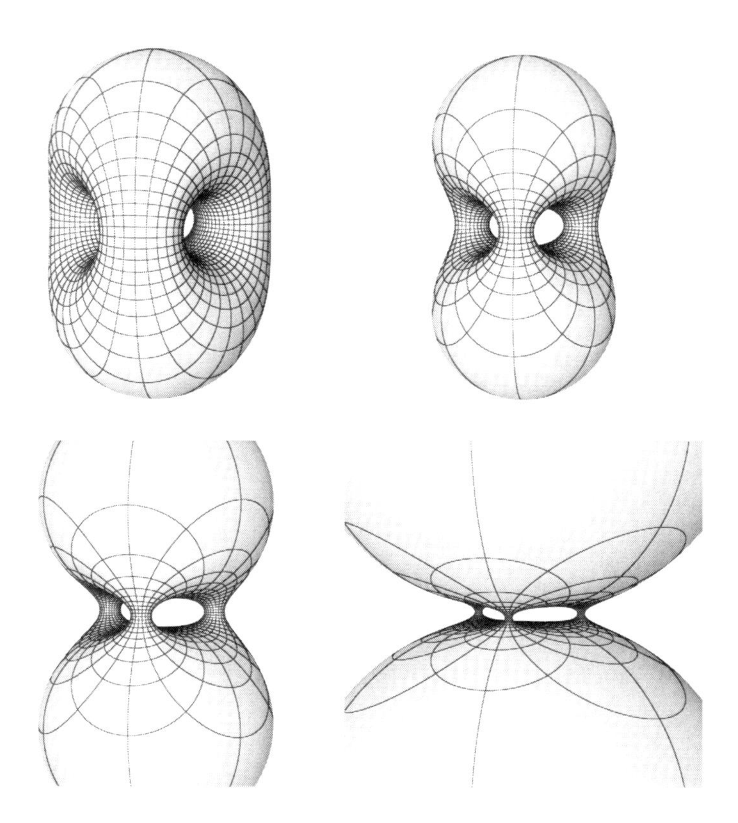

FIGURE 5.4: Family of embedded constant mean curvature surfaces of genus two in the 3-sphere starting at the Lawson surface $\xi_{2,1}$ and converging to a double cover of an equatorial 2-sphere as the conformal type degenerates [32]. The Willmore energy $\mathcal{W}$ initially increases to a maximum above 8π and then decreases to the limiting value 8π (see Figure 5.5).

Willmore energy with $\mathcal{W}(\xi_{g,1}) < 8\pi$ starting at the Clifford torus $\xi_{1,1}$ with Willmore energy $\mathcal{W}(\xi_{1,1}) = 2\pi^2$. The surfaces of Platonic symmetries have larger energies. Numerical experiments, using an energy decreasing flow [34], suggest that Lawson's minimal surfaces $\xi_{g,1}$ are the minimizers of $\mathcal{W}$ over compact surfaces of fixed genus g.

In analogy to the genus one case, where the constrained minimizers in rectangular conformal classes are conjectured to have constant mean curvature (see the 2-lobe Conjecture 1), one could expect such behavior for specific conformal types also for higher genus surfaces. Recently progress has been made in our understanding of the integrable systems approach to higher genus constant mean curvature surfaces [31], [27]. Starting from Lawson's minimal

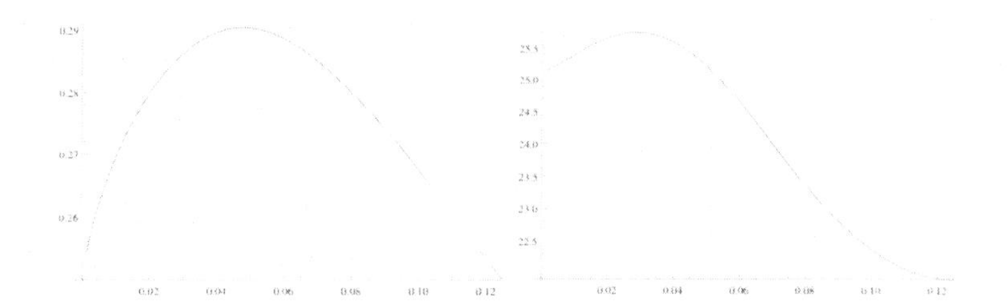

FIGURE 5.5: Graphs of the mean curvature H (left) and Willmore energy $\mathcal{W}$ (right) along the family of genus two constant mean curvature surfaces (Figure 5.4) deforming Lawson's minimal surface $\xi_{g,1}$. The horizontal axis measures the rectangular conformal type, starting at the square structure for $\xi_{g,1}$ on the right and degenerating to the twice covered equatorial 2-sphere at the origin.

surface $\xi_{g,1}$ one can deform $\xi_{g,1}$ by changing the value of the (constant) mean curvature to obtain an (experimentally constructed) 1-parameter family of embedded constant mean curvature surfaces (see Figure 5.4) having the symmetries of Lawson's $\xi_{g,1}$ minimal surface. The Willmore energy profile [28] over this family is shown in Figure 5.5. The conformal types of these surfaces are "rectangular": the quotient $\mathbb{P}^1$ of the surface under the cyclic $g+1$ fold symmetry has four branch points arranged in a rectangle. In particular, the results of [42] apply and the existence of a constrained minimizer in those conformal classes for which $\mathcal{W} < 8\pi$ is guaranteed. We also conducted experiments [30] using the conformal Willmore flow [17] corroborating the results of [34] and supporting

Conjecture 6 (Higher genus conjecture). *The constrained minimizer of the Willmore energy for a surface of genus g whose conformal type is "rectangular" is given by the constant mean curvature analogs of Lawson's minimal surfaces $\xi_{g,1}$.*

At this juncture one could start speculating about the corresponding constrained Willmore Lawson and stability conjectures in higher genus. It is known [16] that the space of minimal surfaces in the 3-sphere of fixed genus is compact. A similar result has recently been announced by Meeks and Tinaglia [56] for strongly Alexandrov embedded surfaces (these surfaces extend to an immersion of an embedded compact 3-manifold) of positive constant mean curvature. Thus, if one could prove that such surfaces are isolated, then the correct generalization of Lawson's conjecture (and its constant mean curvature analogue) would be that there are only finitely many (strongly Alexandrov) embedded constant mean curvature surfaces of genus g and fixed mean curvature. They presumably have different conformal types and that would make them the natural candidates for constrained minimizers in those conformal classes. At present time such investigations would be based on very scant evidence due to the lack of generic examples in higher genus. The importance of examples to develop a mathematical theory has been known throughout its history. In our context this has been well expressed in the opening sentence of Lawson's paper [49], which has served as a starting point for many of the investigations discussed in this note: "It is valuable when dealing with a non-linear theory, such as the study of minimal surfaces, to have available a large collection of examples for reference and insight".

Bibliography

[1] B. Andrews, H. Li. *Embedded constant mean curvature tori in the three-sphere*, J. Diff. Geom., **99** (2015), no. 2, 169–189.

[2] M. Babich, A. Bobenko. *Willmore tori with umbilic lines and minimal surfaces in hyperbolic space*, Duke Math. J., **72** (1993), no. 1, 151–185.

[3] M. Bauer, E. Kuwert. *Existence of minimizing Willmore surfaces of prescribed genus*, Int. Math. Res. Not., **10** (2003), 553–576.

[4] W. Blaschke. Vorlesungen über Differentialgeometrie. Vol. 3, Springer-Verlag, Berlin, 1929.

[5] A. Bobenko. *All constant mean curvature tori in $\mathbb{R}^3$, S^3, and H^3 in terms of theta functions*, Math. Ann., **290** (1991), 209–245.

[6] C. Bohle. *Constrained Willmore tori in the 4-sphere*, J. Diff. Geom., **86** (2010), 71–131.

[7] C. Bohle, K. Leschke, F. Pedit, U. Pinkall. *Conformal Maps from a 2-Torus to the $4-$Sphere*, J. Reine Angew. Math., **671** (2012), pp 1–30.

[8] C. Bohle, F. Pedit, U. Pinkall. *The spectral curve of a quaternionic holomorphic line bundle over a 2-torus*, Manuscripta Math., **130** (2009), no. 3, 311–352.

[9] C. Bohle, I. Taimanov. *Euclidean minimal tori with planar ends and elliptic solitons*, IMRN (2014), **113**, 1–26.

[10] S. Brendle. *Embedded minimal tori in S^3 and the Lawson conjecture*, Acta Mathematica, **211** (2013), 177-190.

[11] R. Bryant. *A duality theorem for Willmore surfaces*, J. Diff. Geom., **20** (1984), 23–53.

[12] R. Bryant. *Surfaces in conformal geometry*: The mathematical heritage of Hermann Weyl. Proceedings of Symposia in Pure Mathematics, **48** (1988).

[13] F. Burstall, D. Ferus, F. Pedit, U. Pinkall. *Harmonic tori in symmetric spaces and commuting Hamiltonian systems on loop algebras,* Ann. of Math., **138** (1993), 173–212.

[14] F. Burstall, A. Quintino. *Dressing transformations of constrained Willmore surfaces,* Comm. Anal. Geom., **22** (2014), 469–518.

[15] E. Calabi. *Minimal immersions of surfaces in Euclidean spheres,* J. Differential Geom. **1**, no. 1-2 (1967), 111–125.

[16] H.I. Choi, R. Schoen. *The space of minimal embeddings of a surface into a three- dimensional manifold of positive Ricci curvature*, Invent. Math. **81** (1985), 387–394.

[17] K. Crane, U. Pinkall und P. Schröder. *Robust Fairing via Conformal Curvature Flow*, ACM Trans. Graph., **32** (2013), no. 4.

[18] J. Dorfmeister, F. Pedit, H. Wu. *Weierstrass type representation of harmonic maps into symmetric spaces,* Comm. Anal. Geom., **6** (1998), 633–668.

[19] J. Dorfmeister, P. Wang. *Willmore surfaces in spheres via loop groups I: generic cases and some examples*, preprint: arXiv:1301.2756v4, submitted.

[20] N. Ejiri. *Willmore surfaces with a duality in S^n*, Proc. London Math. Soc., **57**(2) (1988), 383–416.

[21] D. Ferus, K. Leschke, F. Pedit, U. Pinkall. *Quaternionic holomorphic geometry: Plücker formula, Dirac eigenvalue estimates and energy estimates of harmonic 2-tori,* Invent. Math., **146** (3), 507–593.

[22] D. Ferus, F. Pedit. *S^1-equivariant minimal tori in S^4 and S^1-equivariant Willmore tori in S^3*, Math. Z., **204** (1990), no. 2, 269–282.

[23] D. Ferus, F. Pedit, U. Pinkall, I. Sterling. *Minimal tori in S^4*, J. Reine Angew. Math., **492** (1992), 1–47.

[24] L. Heller. *Equivariant constrained Willmore tori in the* 3*-sphere,* Math. Z., **278** (2014), no. 3, 955–977.

[25] L. Heller. *Constrained Willmore and CMC tori in the* 3*-sphere,* Diff. Geom. Appl., **40** (2015), 232–242.

[26] L. Heller. *Constrained Willmore tori and elastic curves in 2-dimensional space forms,* Comm. Anal. Geom., **22**, no. 2 (2014), 343–369.

[27] L. Heller, S. Heller, N. Schmitt. *Navigating the space of symmetric CMC surfaces*, preprint: arxiv: 1501.01929.

[28] L. Heller, S. Heller, N. Schmitt. *Exploring the space of symmetric CMC surfaces*, preprint: arXiv:1503.07838.

[29] L. Heller, Ch. B. Ndiaye. *First explicit constrained Willmore minimizers of non-rectangular conformal class.* Preprint 2016.

[30] L. Heller, F. Pedit, U. Pinkall, N. Schmitt, U.Wagner. *Conformal Willmore flows on 2-tori in* $\mathbb{R}^3$, in preparation.

[31] S. Heller. *A spectral curve approach to Lawson symmetric CMC surfaces of genus* 2, Math. Ann., Volume 360, Issue 3 (2014), 607–652.

[32] S. Heller, N. Schmitt. *Deformations of symmetric CMC surfaces in the 3-sphere*, Exp. Math., **24** (2015), no. 1, 65–75.

[33] N. Hitchin. *Harmonic maps from a 2-torus to the 3-sphere*, J. Diff. Geom., **31** (1990), no. 3, 627–710.

[34] L. Hsu, R. Kusner, Rob, J. Sullivan. *Minimizing the squared mean curvature integral for surfaces in space forms*, Experiment. Math., **1** (1992), no. 3, 191–207.

[35] U. Hertrich-Jeromin, U. Pinkall. *Ein Beweis der Willmoreschen Vermutung für Kanaltori*, J. Reine Angew. Math., **430**, 21–34.

[36] N. Kapouleas. *Constant mean curvature surfaces constructed by fusing Wente tori*, Invent. Math., **119** (1995), no. 3, 443–518.

[37] N. Kapouleas. *Compact constant mean curvature surfaces in Euclidean three-space*, J. Diff. Geom., **33** (1991), no. 3, 683–715.

[38] H. Karcher, U. Pinkall, I. Sterling. *New minimal surfaces in* S^3, J. Diff. Geom., **28** (1988), no. 2, 169–185.

[39] M. Kilian, M. Schmidt, N. Schmitt. *Flows of constant mean curvature tori in the 3-sphere: the equivariant case*, J. Reine Angew. Math., **707** (2015), 45–86.

[40] M. Kilian, M. Schmidt, N. Schmitt. *On stability of equivariant minimal tori in the 3-sphere*, J. Geom. Phys., **85** (2014), 171–176.

[41] R. Kusner. *Comparison surfaces for the Willmore problem*, Pacific J. Math., **138** (1989), no. 2, 317–345.

[42] E. Kuwert, Y. X. Li. $W^{2,2}$*-conformal immersions of a closed Riemann surface into* $\mathbb{R}^n$, Comm. Anal. Geom., **20** (2012), no. 2, 313–340.

[43] E. Kuwert, J. Lorenz. *On the stability of the CMC Clifford tori as constrained Willmore surfaces,* Annals of Global Analysis and Geometry, **44** (2012), 1–20.

[44] E. Kuwert, R. Schätzle.*Removability of point singularities of Willmore surfaces,* Ann. of Math., **160** (2004), 315–357.

[45] E. Kuwert, R. Schätzle. *Closed surfaces with bounds on their Willmore energy,* Annali della Scuola Normale Superiore di Pisa - Classe di Scienze, **11**, (2012), 605–634.

[46] E. Kuwert, R. Schätzle. *Minimizers of the Willmore functional under fixed conformal class*, J. Diff. Geom., **93**, (2013), 471–530.

[47] J. Langer, D. Singer. *The total squared curvature of closed curves,* J. Diff. Geom. **20** (1984), no. 1, 1–22.

[48] T. Lamm, R. Schätzle. *Conformal Willmore Tori in* $\mathbb{R}^4$, to appear in J. Reine Angew. Math.

[49] B. Lawson. *Complete Minimal Surfaces in* S^3, Ann. of Math., **92** (1970), no. 3, 335–374.

[50] K. Leschke, F. Pedit, U. Pinkall. *Willmore tori in the 4-sphere with nontrivial normal bundle,* Math. Ann., **332** (2005), no. 2, 381–394.

[51] P. Li, S. T. Yau. *A new conformal invariant and its applications to the Willmore conjecture and the first eigenvalue of compact surfaces,* Invent. Math., **69** (1982), 269–291.

[52] X. Ma. *Adjoint transform of Willmore surfaces in* S^n, Manuscripta Math., **120** (2006), no. 2, 163–179.

[53] X. Ma, P. Wang. *Willmore 2-Spheres in* S^n*: A survey,* Geometry and Topology of Manifolds. Springer Japan, 2016. 211-233.

[54] X. Ma, C. Wang, P. Wang. *Classification of Willmore 2-spheres in the 5-dimensional sphere*, preprint: arXiv:1409.2427, to appear in J. Diff. Geom.

[55] F. Marques, A. Neves. *Min-Max theory and the Willmore conjecture,* Ann. of Math., **179** (2014), 683–782.

[56] W. Meeks III, G. Tinaglia. Private communication (2017).

[57] S. Montiel, A. Ros. *Minimal immersion of surfaces by the first eigenfunction and conformal area,* Invent. Math., **83** (1986), 153–166.

[58] S. Montiel. *Willmore two spheres in the four-sphere,* Trans. Amer. Math. Soc., **352** (2000), no. 10, 4469–4486.

[59] Ch. B. Ndiaye, R. Schätzle. *New examples of conformally constrained Willmore minimizers of explicit type*, Adv. Calc., **8** (2015), no. 4, 291–319.

[60] Ch. B. Ndiaye, R. Schätzle. *Explicit conformally constrained Willmore minimizers in arbitrary codimension*, Calc. Var. Partial Differential Equations, **51** (2014), no. 1-2, 291–314.

[61] U. Pinkall. *Hopf Tori in* S^3, Invent. Math., **81**, (1985), no. 2, 379–386.

[62] Pinkall and Sterling. *On the classification of constant mean curvature tori*, Ann. of Math., **130** (1989), 407–451.

[63] J. Richter. *Conformal maps of a Riemann surface into the space of Quaternions*, Doctoral thesis at TU Berlin 1997.

[64] T. Rivière. *Analysis aspects of the Willmore functional*, Invent. Math., **174** (2008), no. 1, 1–45.

[65] A. Ros. *The Willmore conjecture in the real projective space*, Math. Research Letters, **6** (1999), 487–493.

[66] R. Schätzle. *Conformally constrained Willmore immersions*, Adv. Calc. Var., **6** (2013), no. 4, 375–390.

[67] M. Schmidt. *A Proof of the Willmore Conjecture*, arXiv:math/0203224.

[68] M. Schmidt. *Existence of minimizing Willmore surfaces of prescribed conformal class*, arXiv:math/0403301.

[69] L. Simon. *Existence of surfaces minimizing the Willmore Functional*, Comm. Anal. Geom., **1**, (1993), 281–326.

[70] G. Thomsen. *Über konforme Geometrie I: Grundlagen der konformen Flächentheorie*, Hamb. Math. Abh., **3** (1923), 31–56.

[71] T. Willmore. *Note on embedded surfaces*, An. Stiint. Univ. "Al. I. Cuza" Iasi Sect. I a Mat., **11**, (1965), 493–496.

Index